Father
Benito Viñes

Father
Benito Viñes

The 19th-Century Life and Contributions of a Cuban Hurricane Observer and Scientist

Luis E. Ramos Guadalupe
Translated by Oswaldo García

AMERICAN METEOROLOGICAL SOCIETY

Published by the American Meteorological Society
45 Beacon Street, Boston, Massachusetts 02108

For more AMS Books, see http://bookstore.ametsoc.org.

The mission of the American Meteorological Society is to advance the atmospheric and related sciences, technologies, applications, and services for the benefit of society. Founded in 1919, the AMS has a membership of more than 13,000 and represents the premier scientific and professional society serving the atmospheric and related sciences. Additional information regarding society activities and membership can be found at www.ametsoc.org.

Library of Congress Cataloging-in-Publication Data is available at www.ametsoc.org.

Contents

Luis E. Ramos Guadalupe

Foreword

"I reached that part of the building by climbing stairs on the verge of collapse, crossing rooms with sagging floors and crumbling walls, finally reaching the old Observatory. I was alone in that deserted corner of the building, feeling the afternoon breeze blowing through the openings that used to frame windows that looked out on the bay and old Havana. There, blurring my vision, I could almost make out a small man, in black cassock wearing small round spectacles, who without looking at me, said: Can you see those clouds? They are cirrus. I've been waiting for them since yesterday."

The writer is Luis Enrique Ramos Guadalupe, noted Cuban historian and geographer, who has written a fascinating and engaging biography of one of the most famous hurricane forecasters and meteorologists of the late nineteenth century, Father Benito Viñes (1837–1893). Arriving in Havana from France in March 1870, Viñes, a slightly built Jesuit scholar with a keen interest in tropical cyclones, established himself as Director of the Belén Observatory and immediately set about to improve the observational and scientific basis for hurricane forecasting in Cuba. Over the next 23 years, he established himself as a keen observer and astute forecaster of hurricanes, in a time when there were no observations other than a few surface reports sent by telegraph and those that his own eyes provided.

Luis Enrique Ramos writes a lively account of Father Viñes' struggles and successes in establishing himself as a world expert on hurricanes. Born in Havana in 1955, Luis is a geographer by background, and has worked for the last 20 years in the National Museum of Natural History, the National Museum of the History of Sciences "Carlos J. Finlay" and now in the Academy of Sciences of Cuba. He is a sought-after speaker in Cuba, Mexico, Brazil, and Spain and has published over fifty articles, many on the history of meteorology. He has received the Benito Viñes Award of the Meteorological Society of Cuba, which he helped found in 1992, the *Reason for Being* award from the Alejo Carpentier Foundation, the *Gerado Castellanos* award by the Municipal Culturem and the Museum of Guanabacoa, and *Memory* by the Fundación Pable de la Torriente Brau.

Father Benito Viñes: The 19th-Century Life and Contributions of a Cuban Hurricane Observer and Scientist, was translated from Spanish into English by my friend and colleague Professor Oswaldo Garcia from San Francisco State University. Oswaldo accompanied me on my first trip to Cuba in March 2007, a trip where I met Luis Enrique Ramos for the first time. Oswaldo was born in Havana in 1947 and projects an infectious enthusiasm for the Cuban people, science, and a stronger relationship between Cuba and the United States. I readily agreed to help him with his translation, but in truth very little help was needed.

Matching Oswaldo's enthusiasm for science, history, and culture, Luis Enrique Ramos has become a treasured friend and colleague, and I was delighted to respond to his request to introduce his book. I heartily recommend this book to everyone, whether a historian, scientist, or someone just interested in a fascinating tale of a dedicated priest, scientist, and forecaster who wrote:

"I do not desire any reward, other than the one I expect from God, than being useful to my brethren and to contribute in some fashion to the advancement of science and the welfare of humanity. Benito Viñes

—Richard A. Anthes
President Emeritus
University Corporation for Atmospheric Research
Boulder, Colorado

Translator's Note

A set of fortuitous and unexpected circumstances led to my first visit to Cuba in 2005 after a 45-year absence. During that brief stay in Havana, I had the opportunity to attend a meeting of the Cuban Meteorological Society, now headquartered across the street from the school I attended as a boy. There I met Luis Ramos, and was immediately impressed by his encyclopedic knowledge of the history of meteorology in Cuba and I first learned of the long and almost uninterrupted collaborations between Cuban and American meteorologists that go back well over a century. That initial trip to Havana led to other visits where I had the privilege of accompanying Rick Anthes and other American meteorologists to Cuba. This book is a tangible outcome of these visits.

The Spanish edition of this book was published in Cuba in 1993, before the Internet was widely available in that country. Many information resources that were not available then are now easily found online, including some of the original texts written by pioneers in the field of meteorology with which Father Viñes was well acquainted. Keeping in mind an English language readership, I took the opportunity to add many footnotes to this edition of the text, including biographical notes about nineteenth-century meteorologists and some Cuban historical figures of the time. I hope these footnotes will be of help to some readers.

While preparing the translation of this text, I spent many hours at the New York Public Library obtaining information about the sinking of the *Steamship Liberty* during the hurricane of September 1876, which allowed me to supplement the author's original description of that event in Chapter 1 of this book. The author also sent me more information about the astronomical observations made by Fr. Viñes which I was able to incorporate into the English text.

I thank Rick Anthes for his encouragement and patience throughout this process, and for the many hours he spent poring over my drafts. Without his timely help, its publication would not have been possible. My thanks also to Tom Bestor, who reviewed the manuscript and made useful suggestions, and to the staff of the New York Public Library who provided me access to material that would have been very difficult to obtain otherwise.

—*Oswaldo García*
Professor of Meteorology
Department of Earth & Climate Sciences
San Francisco State University
San Francisco, California

Acknowledgments

The author wishes to extend his sincere thanks to the following people who made this book possible:

To Dr. Richard Anthes for his friendship and the efforts he made to facilitate the translation and publication of this work in the United States of America, and in particular for his close reading of the English language manuscript and the useful suggestions for making it better.

To Dr. Oswaldo García for his friendship and tireless work in translating the original Spanish text and the valuable additions he made to the text, which helped update and enrich its content.

To Sarah Jane Shangraw and Beth Dayton of the American Meteorological Society, whose professionalism and patience helped sustain me during the long process of publication.

To Gerard Martí i Jarque, from the native land of Father Viñes, who generously made it possible for me to obtain many of the original documents I have used in this work.

To Dr. Mario Rodríguez-Ramírez, in memoriam, for his help in locating and facilitating my access to the historical archives of the Observatory of the College of Belén in Havana.

To Dr. Ramón Pérez-Suárez, researcher at the Cuban Institute of Meteorology, for his careful editing of the original text in Spanish.

To Reverend Father Felicísimo, S. J., in memoriam, of the Sacred Heart Parish in Havana, for opening the door to my initial work on this book, years ago.

To Lázaro Calderín-Vega, who made the first edition of this work in Cuba possible.

And thanks also to Adalberto Sarría-Savory, for his unconditional support and encouragement in all my research endeavors.

This book is dedicated to the memory of Benito Viñes-Martorell, S.J., and to the Cuban and American meteorologists who have been instrumental in promoting collaborations between our two countries since 1855.

Introduction

A thin man, dressed in black clerical robes, descends the gang-plank leading down to the pier from a ship flying the French flag. The scene unfolds on the "Light Wharf" in nineteenth-century Havana. The man who has just landed is young, not quite 33 yet, but already wearing small spectacles. His demeanor exudes a peculiar combination of weariness and absent-mindedness, more typical of those who have led a long life of the mind. He, of course, has had his share of education, having been recently ordained in the Jesuit order, and his arrival is anticipated at the Royal College of Belén of the Company of Jesus in Havana, where he is to head its observatory. His name is Benito Viñes, although in a few months he will simply be known as "Father Viñes," researcher of rotating tropical storms. By then, the weariness he had initially projected—a result of the long voyage from Europe—will have disappeared, and his tireless, energetic nature fully revealed.

Research in meteorological topics was not new to Cuba. Back in 1847, don Desiderio Herrera Cabrera[1] had published his Memoirs

---

[1] Desiderio Herrera Cabrera (1792–1856) was a Cuban educator, mathematician, and land surveyor. His book on hurricanes was the first to take an empirical, descriptive approach to the study of hurricanes affecting the island.

on the *Hurricanes of the Island of Cuba,* and Andrés Poey y Aguirre[2] two years later was already requesting weather reports from various parts of the island with the ultimate objective of establishing a network of meteorological observing stations in Cuba (Ortiz 1979).

After several failed attempts, Poey tried again, pleading on the pages of the daily newspaper *Diario de la Marina (Diario de la Marina 1850)* for the Spanish colonial authorities to make funds available to establish such a network. Facing an uncertain funding scenario, Poey built an observatory in his own home, where he conducted weather observations during the course of a year before he moved to the United States.

Thus things remained when in 1852 Don José María de la Torre[3] petitioned the Vice Royal Protector of Public Instruction to establish an observatory in Havana focused on meteorological observations. Finally, after many unsuccessful attempts, the efforts bore fruit when on December 8, 1860 the Spanish Queen Isabel II issued a royal decree ordering the establishment of an observatory in Cuba (Archivo Nacional de Cuba, 1861). With Andrés Poey himself at the helm, the Physical-Meteorological Observatory was established the following year on the grounds of the Royal Economic Society, on Dragones Street in Havana. Poey was sacked on February 26, 1869, after being accused of "negligence."

---

[2] Andrés Poey (Havana, 1825–Paris, 1919) was a pioneer of Cuban meteorology, a founding member of Havana's Royal Academy of Medical, Physical and Natural Sciences, and first director of the Physical-Meteorological Observatory of Havana—the first meteorological service established on the island—established in 1861. He was a member of the French Meteorological Society and author of more than 100 publications in the fields of meteorology, archaeology, and positivist philosophy. At the request of the French government, he established a meteorological observatory in Mexico during the reign of Maximilian.

[3] Jose María de la Torre was a Cuban geographer. Among his accomplishment was the creation of a map of the island, published in 1875.

After Poey's departure, the Observatory moved to the site of the *Escuelas Profesionales* (Professional Schools) and later to the campus of the University of Havana in 1887. In the meantime, the College of Belén's observatory, which had been established back in the mid-1850s, had begun archiving daily weather observations. The man who was now to lead this observatory came with a two-pronged agenda: first, to develop a set of principles dedicated to forecasting the movement of hurricanes by systematically observing the movement of clouds, and second, to establish a network of weather observations in the Caribbean Basin.

This book is dedicated to the memory of this remarkable man. There are many aspects of this man that have remained unclear or have been heretofore ignored. This work, done in preparation of the centennial of his death in 1993, has proven difficult, as firsthand accounts of his life are lacking. They are based on three primary sources: his record as Fellow of the Royal Academy of Medical, Physical and Natural Sciences of Havana, the Annals of that institution, and a carefully tended archive of newspaper clippings that was maintained by the Colegio de Belén itself.

From those sources and from an intensive search of publications of that era that can be found at the National Library and the Institute of Literature and Linguistics, I have compiled this historical essay, both biographical and bibliographical, which is both a historical research and a retrospective look of the work of this learned Jesuit priest.

An important part of the material used in this work is the author's personal library, and I am responsible for the sketches of the maps that appear as a complement of the text. May this work serve as a modest homage to the selfless scientist, the tenacious researcher, who never rested until the day's work was done, and whose lifespan was not long enough to accomplish all that he set out to do.

I conclude this Introduction with quotes from his successors in the Observatory:

> The name of Fr. Viñes will always be spoken with veneration and affection, and his memory will always be linked to memorable events and achievements, and will always be kept alive among those that knew him. His wise and practical science will continue to provide incalculable service to mankind. (Gangoiti, 1895)

> ...and regarding the accuracy of Fr. Viñes' forecasts, it is hard to believe the depth of knowledge and accuracy that he was able to achieve in this area. It seemed that this man was able to read in the clouds the intentions of the hurricane. (Gutiérrez-Lanza, 1904, p. 99)

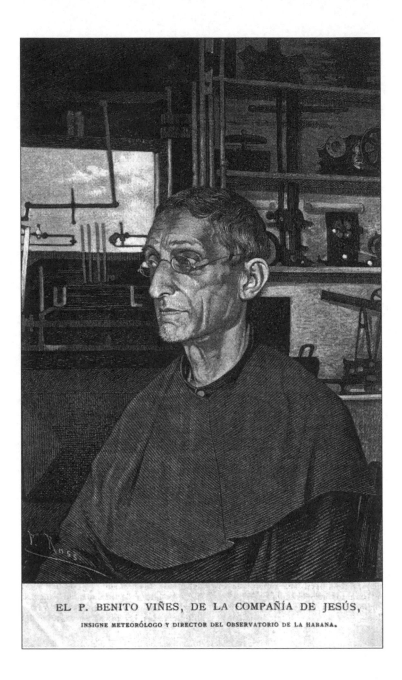

EL P. BENITO VIÑES, DE LA COMPAÑÍA DE JESÚS,

INSIGNE METEORÓLOGO Y DIRECTOR DEL OBSERVATORIO DE LA HABANA.

First Years in Cuba 1

I do not desire any other reward, besides the one I expect from God, than being useful to my brethren and to contribute in some fashion to the advancement of science and the welfare of humanity. —Benito Viñes (Gutiérrez-Lanza 1942, p. 137)

Never had Matanzas felt the effects of a disaster of such magnitude, nor had it experienced such ruin and desolation... Once the hurricane began, it mowed through everything it encountered, and entire families—destitute, naked, hungry and without shelter—turned their eyes heavenward seeking safe shelter to ride out the furious winds. But to no avail: the storm only got worse; the two rivers that divide Matanzas jumped their banks and swept whole houses that had already collapsed from the force of the wind into the swift current. The multitudes of people fleeing from the scene were unable to escape the torrent that engulfed them, and perished in the high waves. (Aurora del Yumurí 1870)

We walked till one in the morning when the San Juan and Yumurí rivers flooded, which is when the horror began...to get an idea of the magnitude of the flooding, it is enough to know that the water level surpassed the roofs of the houses in the lower part of town, the current carrying away those that were taking refuge in them...according to police reports, two thousand people are believed to have perished. At the San Luis station several people took refuge on a rooftop and fashioned an improvised raft with parts of the support beams till the current carried it away...at six in the morning the raft was carried downstream toward the bay, its occupants pleading for mercy. There were some forty people on that raft, which was last seen engulfed by a wave. (Diario de la Marina 1870)

These were undoubtedly the first accounts of a hurricane Benito Viñes had ever read. Both were published in newspapers of the time, copies of which are in the Colegio de Belén archives. The first was published in *La Aurora del Yumurí* of Matanzas, and the latter in the *Diario de la Marina* of Havana. Both articles refer to the terrible hurricane that hit Matanzas on October 7 and 8, 1870.

Viñes had arrived in Havana seven months before, equipped with only a theoretical knowledge of tropical meteorology, and most likely with the strong desire to obtain a first-hand knowledge of these storms called cyclones or hurricanes, typical of these regions, storms he knew about only from his readings.

One can only imagine the strong impression those accounts must have made on him, the inevitable comparisons he must have made between the writings of William Redfield[4] and Henry

---

[4] William Charles Redfield (1789–1857) was an American scientist, founder and first president of the American Association for the Advancement of Science. He is credited as being the first to describe the cyclonic circulation around the center of tropical storms. However, he did not identify the convergent nature of surface winds in tropical storms, a fact Viñes later described in his published work.

Piddington,[5] and the way events unfolded right in his backyard, with the tremendous consequences they had on the society and economic life of the country in which he had just arrived.

This hurricane, coming right on the heels of his arrival in Cuba, was symbolic of the arduous and complex work that lay ahead of him, and of the urgency of doing his best to mitigate the impacts of these natural disasters. To that effect, he invoked the help of God and of Our Lady of Charity of Cobre, patron of Cuba, whose statue was found floating in rough seas by fishermen during a similar storm in 1604. It would not have escaped his notice that an inscription on the statue read: *In fluctibus maris ambulavit*, or "She who walks on stormy seas," and that the symbolism and content were no mere coincidence.[6]

When did Fr. Viñes arrive in Cuba? According to his most prominent biographer, Fr. Mariano Gutiérrez-Lanza,[7] Benito Viñes arrived in Cuba from France in March 1870. In his own words: *"As soon as Fr. Viñes took over as director of the Observatory he dedicated himself to his task with passion and perseverance befitting his genius"* (Gutiérrez-Lanza 1904, p. 57).

---

[5] Henry Piddington (1792–1858) was a British official in Colonial India, who served as President of the Marine Court of Inquiry in Calcutta. He is credited with coining the word "cyclone" to describe rotating storms.

[6] The feast of Our Lady of Charity of El Cobre (Nuestra Señora de la Caridad del Cobre) is celebrated on September 8, the month with the highest frequency of hurricanes in the Atlantic.

[7] Mariano Gutiérrez-Lanza (Spain, 1865–Havana, 1943) was a Spanish Jesuit who made contributions to the field of tropical meteorology. He arrived in Cuba in 1896, studied at Georgetown University, and worked for a time at the U.S. Naval Observatory. He took over as Director of the Belén Observatory in 1925, where he became well known as a hurricane forecaster.

Gutiérrez-Lanza gives us a more precise date for Fr. Viñes' arrival in a monograph he wrote in the 1940s entitled "Father Viñes, S.J.", part of an anthology of talks given by various authors on the topic of "Prominent Cuban Scientific Researchers." There he tells us:

> *The Colegio de Belén log of March 7, 1870 states: This morning, Fathers Miguel Mora and Benito Viñes arrived on the French steamer; the former will be in charge of the pulpit and the latter will be running the meteorological observatory.* (Gutiérrez-Lanza 1942, p. 112)

In another work by the same author one can, however, read the following "...Fr. Viñes, from the day he arrived, March 4th 1870, took on the daily work of the observatory with characteristic energy" (Gutiérrez-Lanza 1936, p. 12).

One can see that conflicting arrival dates are given, both in March 1870. The few authors who have written on the subject use one date or the other, as they all base their assertions on the work of Gutiérrez-Lanza.

To settle this discrepancy we attempted to verify the manifests of ships arriving and departing the port of Havana for the month of March 1870, as published in the known newspaper of record of the time, the *Diario de la Marina*. Copies of that newspaper for 1870 are kept at both the José Martí National Library and the Institute of Literature and Linguistics, but unfortunately they have deteriorated to such a state as to prevent their inspection and handling.

However, with enormous care and patience we were able to piece together some pages of the 4 March 1870 issue of the *Diario de la Marina* from the collections of the National Library. After assembling pieces of the jigsaw puzzle we reconstructed the following note:

The French steamer "France" of the Saint Nazarie lines arrived today… [followed by illegible lines and ending with] …The French steamer "France" will depart at noon tomorrow, the 5th, for Veracruz. (Diario de la Marina 1870)

Although we were unable to reconstruct the passenger roster corresponding to this ship's arrival notice, due to the crumbling pages in which the information was found, the arrival date is clear. This confirmed for us that one of the two dates given by Gutiérrez-Lanza corresponds to the arrival in Havana of the ship from France, where Viñes had just completed his education.

As far as the other date cited by Gutiérrez-Lanza in the Colegio de Belén log, it is possible that the March 7 date corresponds to the day in which the information was entered into the log, and not to his actual arrival date in Havana. It is plausible that three days may have transpired between his arrival and the date of the log entry. It is worth noting that the lack of personnel in the College, as will be discussed later, could have accounted for this delay. We have searched in vain for this log, and it is quite possible that it no longer exists.

By sheer historical coincidence, March 4, 1870 was also the date on which a Spanish military tribunal sentenced José Julián Martí y Perez,[8] a Havana youth of seventeen, to six years in prison for his political activities.

One can assume that immediately upon his arrival in Havana, Viñes began discharging his duties as director of the Observatory, since he had been appointed to this office before leaving Europe (Gutiérrez-Lanza 1936, p. 1).

---

[8] Full name of José Martí (1853–1895), who was the leading figure in Cuba's struggle for independence from Spain in the late 1800s. He was also an important figure in Latin American literature as an essayist, journalist, revolutionary philosopher, and translator.

Upon his arrival, Viñes met the rector of the College, Father Andrés García Rivas, S.J., who held that position from 1868 to 1874. Fr. García Rivas would have given Viñes his charge and assigned him a small room in the convent that would become Fr. Viñes' home in Cuba.

FIGURE 1. Undated photograph of Belén College, from approximately 1890.

It must have come as a shock to Viñes to see firsthand the state of disarray in which the Observatory's instruments and records had been kept, as we can read in Gutiérrez-Lanza's characteristically sober accounting of the College's internal affairs:

> *The Observatory continued operation for three years: 1867 and 1868 under the direction of Fr. Pons, with Fr. Joaquín Rollán serving as associate, and 1869 under Fr. José María Vélez by himself…The situation at the Observatory had become extremely critical. With the Observatory in such a sorry state, the indomitable Fr. Viñes took charge of it.* (Gutiérrez-Lanza 1904, p. 20)

In fact, the Observatory had been suffering a crisis, the worst since its founding in 1857. After ten years of operation, in the period between 1867 and 1869, it was struggling for its very survival. This coincided with the time when the Jesuits were expelled from Spain after the Revolution of 1868, which had a marked radical and anti-clerical tone.

Thus Fr. Viñes became the Observatory's seventh director, finding it almost abandoned for lack of personnel and resources. One must assume that rigorous and methodical observations were no longer being performed and necessary tasks were delayed or neglected, although not completely abandoned.

In those days the Observatory occupied a humble corner of the College, not due to its lack of importance, but by virtue of the dilapidated quarters it occupied on the third floor of the building. This is what Viñes himself wrote in a press release on the occasion of the 1870 hurricane, which is likely the first weather report ever sent to Havana newspapers:

> ...Magnetic instruments indicated only a small perturbation until mid-day on the 19th, most notably in the bifibilar magnetometer, which near 11 PM began oscillating in an irregular fashion. After that time we cannot give a further account, since we prudently abandoned the Observatory for a safer place. And in retrospect it was a wise decision, since at 4:15 the wind strengthened, and the zinc cover that protected the roof opening of the Observatory was blown 6 to 8 meters away...
> (Diario de la Marina 1870)

The Observatory sustained serious damage during this event, evidence of its precarious state at the time. In a few weeks, however, his energy, steadfastness, and work ethic had put the work of that branch of the college on the right track, turning it in short order into one of the principal observatories located in the tropical regions of the world at that time.

By then, three urgent tasks had risen to the top of his agenda: First was to re-establish regular meteorological observations, second to quality check and calculate climatological data from weather observations recorded at the site, and third to initiate research in tropical meteorology and the causes of the weather observed in Havana.

> *As soon as he took over, his skilled hand became apparent. Routine tasks finally received the attention and care they needed, and making use of the data gathered in the 12 years [sic] since the foundation of the Observatory, he began coordinating observations of atmospheric phenomena and developing preliminary research that led, years later, to his glorious achievements in the understanding of cyclones.* (Gutiérrez-Lanza 1904, p. 20)

It was still premature for him to delve into the area of forecasting. In general, authors that have written about Fr. Viñes are few in number and have not established the exact date when his hurricane forecasts began. Fr. Gutiérrez-Lanza writes:

> *We cannot ascertain precisely the year when Fr. Viñes began his cyclone forecasts. The first known public forecasts are for the 11th and 14th of September 1875. But there are reasons to believe that even before these dates Fr. Viñes shared with the press his opinions about the probable trajectory of cyclones, although no data on this have survived...* (Gutiérrez-Lanza 1904, p. 68)

I do not agree completely with Gutiérrez-Lanza's assertion that Viñes likely made forecasts during the 1870-1875 period, since there are strong indications of circumstances that account for the absence of his celebrated "announcements" until 1875. In other words, I believe that Viñes did not issue any hurricane track forecasts until the one in September 1875.

It appears that, lacking other data, Gutiérrez-Lanza must have assumed that, since Viñes had become a well-known scientist as early as 1873, it must have been because of his tropical cyclone forecasts. I beg to disagree with this assumption, based on the following considerations.

There is no doubt that the first tasks that Fr. Viñes must have undertaken starting in 1870, apart from reorganizing the work of the Observatory, were to carefully compile, organize, and evaluate the data that had already been gathered, and to acquire practical experience observing tropical phenomena. It would not have been logical at that stage for him to venture into tropical cyclone forecasting. There is no indication of any attempt by Fr. Viñes to predict the weather in the years 1870 to 1873. During these years he likely would have been developing a clear research methodology, and using it to define the typical behavior of atmospheric phenomena.

During these initial years, as his reputation as a scientist was growing, he focused only on descriptive and statistical studies in the areas of meteorology, geomagnetism, and applied physical science. The results of this work appeared in the form of monographs, which he presented at the top scientific forum of the country: the *Academia Real de Ciencias Médicas, Físicas y Naturales* (Royal Academy of Medical, Physical and Natural Sciences) of Havana.

This institution was founded by Royal Decree in 1860 during the reign of Isabel II of Spain, at the urging of a group of Cuban physicians and intellectuals. It was the first institution of this type in Latin America. Its primary mission was to serve the colonial government of the island (headed by the Governor General) in an advisory role in matters primarily related to medical science, although it had separate sections dedicated to Pharmacy and to the Natural Sciences, the latter being the one to which Viñes belonged. Records of these works can be found in the historical archives of the Cuban Academy of Sciences (ACC in its Spanish acronym), the successor institution to the Royal Academy, which continues its work to this day.

The oldest work by Viñes that can be found in these archives is entitled "Magnetic perturbations and the Aurora Borealis of the 4th of February 1872," which appeared in Volume 9 of the Annals of the Academy for that year (*Anales de la Real Academia* 1872, p. 115). That work was followed by "Magnetic perturbations related to other meteorological elements during the months of June-October, 1872" (*Anales de la Real Academia* 1872, p. 240). The latter was published in several parts, through 1873, all in the *Anales de la Real Academia*.

In volume 10 of the *Anales*, covering 1873 and 1874, Viñes published a monograph entitled "Storm of 6 October 1873" and in the same journal, between 1874 and 1875, "Regular and irregular barometric fluctuations in Havana between 1858 and 1871 (Anales de la Real Academia 1875, p.274).

Here we have four published works on meteorology and geomagnetism—one in particular focused on the study of "storms"—that are sufficient to assure the author's good standing in the developing scientific scene of Havana in the 1870s. In addition to these publications, he also wrote a detailed study of the hurricanes of October 1870 (a second hurricane affected westernmost Cuba a few days after the one of October 7 and 8) that appeared in the form of a pamphlet without imprint (information on the publisher and date and location of publication). This undated pamphlet, now kept in the archives of the ACC, may well have been his first publication in Cuba, and is evidence of his keen interest in the study of hurricanes. As far as we know, this is the only surviving copy of this work. It is quite possible that this pamphlet had been included as an appendix in a larger publication that summarized that year's meteorological observations.

Another no less important factor must be kept in mind in explaining why Fr. Viñes likely did not do hurricane forecasts during his first years in Cuba: the relative absence of hurricanes that affected Cuba during the period 1871-1875. This surely explains why no hurricane forecasts were issued by Fr. Viñes during the first half of the 1870s. Gutiérrez-Lanza himself notes this quiet period in his work *Cyclones that have hit the Island of Cuba* (Gutiérrez-Lanza 1934) in which he states:

> 1871—No cyclone
> 1872—No cyclone
> 1873—October 5-6. Weak cyclone
> 1874—No cyclone

These assertions are also supported in the works of the well-known meteorologists Ivan Ray Tannehill and Mario Rodríguez Ramírez[9] (Tannehill 1938; Rodríguez Ramírez 1956).

These were undoubtedly the determining factors that made Fr. Viñes focus on phenomena other than hurricanes during his initial years as head of the Observatory, refraining from issuing any explicit hurricane forecasts until 1875, when the first one appeared in the local press. It is even possible to explain his relative silence on this matter during 1873 when a relatively weak cyclone passed to the northwest of Cuba near Pinar del Río, Cuba's westernmost province, since he harbored doubts as to whether it was a well-defined hurricane. This doubt is evidenced in the article "Temporal del 6 de octubre de 1873", which refers to a "temporal" (storm) and not to a "ciclón" (cyclone) or "huracán" (hurricane), which were terms already in use by the Jesuit priest.

All evidence points to the fact that, even though Viñes did not issue a formal forecast for the 1870 hurricane, the Belén Observatory, under its new director, was already issuing regular press releases about noteworthy meteorological events. Note the following found in the College's press clipping archive:

---

[9] Mario Rodríguez Ramírez (1911–1996) Distinguished Cuban meteorologist who worked on hurricane research and directed a major expansion of the Cuban meteorological observation network in the 1960s.

*We acknowledge the Reverend Fathers of the Colegio de Belén
the honor they provide "La Voz de Cuba" [The Voice of Cuba]
in making available to us the following observations they have
made…* (La Voz de Cuba 1870)

This is clearly a reference to a report of instrumental observations, not of a weather forecast. Although Fr. Viñes is not explicitly mentioned in this newspaper clipping, one can discern his style of writing and preciseness of language. We will consider later the evolution of his scientific ideas and the genesis of his theories. In the following note about the 1870 storm that he sent to a local newspaper, one can discern important elements that characterize his later work:

*Last night's storm, whose effects we are still feeling, began on
the 18th with a wind shift to the ESE, followed by a significant
drop in the barometric column…All this, and the changing appearance of the sky, together with rapidly strengthening winds,
made us think we were witnessing the beginning of a significant
storm, and prompted us to begin making observations every 15
minutes. We paid special attention to the strength of the wind,
which made us suspect that the storm was approaching from
the SSW, and we anxiously awaited the wind shift to S and SW.*

*This hoped-for shift did not occur until 4:30 AM by which
time the barometer reading had dropped spectacularly and
the wind was roaring…*

*The appearance of the sky at that time was striking. Dark
thick clouds were crossing the lower part of the atmosphere
with uneven speed…The barometer which had reached its
minimum by that time, lingered for a while at that level with
nearly imperceptible variations, and then began rising.*

*The most probable consequences that we can deduce from
our observations, informed by the most generally accepted
theories of the laws governing storms are as follows:*

<table>
<tr><td>1st</td><td>*That the storm is of the gyratory type commonly called cyclones*</td></tr>
<tr><td>2nd</td><td>*That the vortex of the hurricane took a path from SSE to NNW*</td></tr>
<tr><td>3rd</td><td>*That it crossed Havana's latitude at approximately 4:15 am, a few leagues to the W of Havana*</td></tr>
<tr><td>4th</td><td>*That from that point on the area of danger began receding from us*</td></tr>
<tr><td>5th</td><td>*That the current direction of the storm path is toward the NNE and that it is probable that it will turn more toward the E*</td></tr>
</table>

The report continues with a table listing observations where the 4:45 AM observations are highlighted. Pressure: 741.3 mm, wind SSW

We have transcribed such a long quote to show how Fr. Viñes tested theory against observations, and was able to specify the particular characteristics of tropical storms. He concluded from his observations of the veering of the wind that the phenomenon he witnessed was a hurricane, and it is noteworthy that he paid special attention to cloud movement. These would later become essential elements of the methodology he applied to the prediction of hurricanes.

The year 1873 marked two important events in Fr. Viñes' life. The first was his election as a "Socio de Mérito" (Fellow) of the Real Academia de Ciencias Medicas Físicas y Naturales (Royal Academy of Medical, Physical and Natural Sciences, henceforth the Royal Academy). The second was his acquisition of Fr. Secchi's meteograph, a multi-purpose instrument that made possible more accurate measurements of meteorological parameters, while at the same time making the process less cumbersome.

We have seen how in the period between 1870 and 1873 the scientific community, and to some extent the general population of Havana, became aware of the work being carried out by the meteorologist-priest. Most assuredly from early on the Royal Academy members were aware of Viñes' background and expertise. A scientific star of first magnitude was emerging in Cuba, an island that was otherwise lacking in comparable talent after Andrés Poey.

Fortunately, the Center for the Study of Science and Technology (ACC in its Spanish acronym) has preserved various documents related to Viñes that pertain to his dealings with the Royal Academy. His academic record is contained in four binders. They are headed by a folio with personal data, a printed form, similar to those still in use today, requiring a lot of tedious work to fill out then as they do now. It lists his name, date and place of birth, date of his accession to the Royal Academy, and date of his "departure" upon his death. This form is followed by a historical document: the hand-written decree by the executive board appointing him Fellow of the Royal Academy, dated February 23, 1873.

At this meeting, celebrated with great solemnity in the main hall of the Royal Academy, it was resolved to present Fr. Viñes with a diploma certifying his status as Fellow of the Royal Academy, a copy of the Academy's bylaws, and a copy of its library rules. A note regarding his appointment as Fellow was hand-delivered to Fr. Viñes the day after. The note is signed by Dr. Antonio Mestre, Secretary of the Academy and reads as follows:

Appointment of Distinguished Fellow Rev. Fr. Benito Viñes, S.J., February 23, 1873

In an executive session meeting held at this Academy yesterday, and following the adoption of a motion put forth by the President of the Academy, I would like to inform you that you were elected by unanimous vote as a Distinguished Fellow of this Institution, in recognition of the great merit your Excellency has acquired though your work in the service of science. I am enclosing a Diploma attesting to this fact, as well as a copy of the Academy's Bylaws and the Library Rules.

Havana, February 24 1873
Secretary of the Academy

In addition to the above referenced letter there is another note from the minutes of the meeting, in memorandum format, also dated February 24, that states:

Rev. Fr. Benito Viñes, S. J.

Elected Distinguished Fellow in executive session on February 23 1873, in accordance with article 20. Diploma attesting to that election was caused to be issued.

Gutiérrez, President
Miranda, Treasurer
Mestre, Secretary

The signatures on this document belong to don Nicolás José Gutiérrez Hernandez, President of the Academy of Sciences, Ramón Luis Miranda, Treasurer, and Dr. Antonio Mestre, Secretary of the Academy.

Next on the binder is a manuscript of the acceptance speech given by Fr. Viñes on the occasion of his induction to the Academy as a Distinguished Fellow. The handwriting on the manuscript does not correspond to that of Viñes and was probably transcribed by a clerk. It is dated March 9, 1873. Because of its historical importance, we reproduce it below in its entirety:

Mr. President
Members of the Academy
Dear Sirs:

Divine Providence, which guides men in their fulfillment of their sacred obligations, has never failed to provide me with a faithful friend to lend his hand and share his wisdom and give me strength to carry on my work.

Indeed, dear Sirs, the kind invitation your distinguished President has extended to me, the motion he made to this distinguished Academy, and the unanimous vote in support of this motion, bestowing me an honor I thought far from my reach and would never have dared to covet, not only fills me with great satisfaction, but serves to give me renewed strength in following the path I have chosen.

The learned men of your noble institution provide me with a model of hard work, erudition, and ongoing contributions to science and society that I pledge to emulate. Who cannot but take comfort from the shade of the stately tree that your Academy represents, or savor the sweet fruit from its branches? Or drink from the waters of wisdom that emanate from its abundant spring? Or bask in the light that radiates from it?

With all the vigor with which I am endowed, and thanks to the great honor you have bestowed on me, I propose to learn from every one of you and benefit greatly from your wisdom and your help. I must henceforth fulfill my obligations to you in gratitude for this great honor. I most gratefully thank you, Mr. President, for the kindness you have shown me, and I thank in no less cordial or respectful way the members of this Academy, who have honored me by welcoming me so warmly.

I also want to take this opportunity to give you thanks in the name of the Society of Jesus, of my mother, my superiors, and of those that preceded me in this capacity at the Observatory. I welcome the opportunity to contribute our observations to your annals, and thank you for the interest so many of you have shown in our observatory, even to the point of contributing your work to our publications. (ACC Historical Archives)

There is no doubt, therefore, that the nomination of Viñes and its unanimous approval were evidence of his growing reputation among his scientific peers, a scant three years after he began his work in Cuba.

Two more documents pertaining to his appointment with the Royal Academy are available. The first is from the Royal Academy to the colonial government (Gobierno Superior de la Isla de Cuba) informing it of Viñes' appointment:

Dear Sir,

In accordance with Article 8 of our Bylaws, I have the honor of informing Your Excellency that the Reverend Father Benito Viñes, S.J., Director of the Meteorological Observatory of the Colegio de Belén, was appointed Distinguished Fellow of this Academy at a session held yesterday.

February 24, 1873 (ACC Archives)

The second, complete with embossed seal, officially acknowledges its receipt:

Office of the Governor of the Island of Cuba
Secretary of the Development Section

His Excellency the Governor of Cuba has been informed of your communication regarding the appointment of Rev. Fr. Benito Viñes, Director of the Meteorological Observatory of the Colegio de Belén as a Distinguished Fellow of your Acad-

emy. I have been directed by His Excellency to inform you of his concurrence with this appointment.

May God keep you.
Havana, March 28 1873

A second important event in the life of Fr. Viñes took place in May 1873, scant weeks after his accession to the Royal Academy. Viñes acquired a most valuable ally for conducting his research with the arrival from Europe of a "meteograph" designed by Fr. Secchi. The meteograph is best described as a collection of instruments housed together. Unfortunately, no documents could be found regarding its acquisition, but given its sophistication and probable expense, it was likely ordered from its manufacturer at least a year in advance. It was probably manufactured in Italy or France. An inscription on its housing reads "E. BRASSART NO. 6". Fr. Gutiérrez-Lanza describes the meteograph in the following terms:

The principle used for measuring the barometric pressure is very ingenious. The tube, made of iron, hangs by its closed end from one end of a scale. The lower, open end is submerged in a container of mercury. The uppermost section of said tube is wider than the rest, and as the mercury level rises or falls in this section, it changes the weight of the tube, causing the scale to tip one way or the other. A pen, fastened to it through a parallelogram mechanism, records the degree of tipping of the scale. Noteworthy to those seeing it for the first time is the recording device for the wind speed. This device has a clockwork mechanism whose escapement[10] is regulated by an electromagnet activated by each turn of the anemometer. A recording pen traces a horizontal line on a rotating drum proportional

---

[10] An escapement is a device in clocks that transfers energy to the timekeeping element and allows the number of its oscillations to be counted.

to the number of turns of the anemometer during a 15-minute period, after which the position of the pen is reset to another position on the same vertical line (Gutiérrez-Lanza 1904).

FIGURE 2. Father Viñes standing by the Belén Observatory meteograph.

This new acquisition was surely welcomed by Viñes, since it not only represented an opportunity to obtain continuous meteorological records, but also allowed observations to be made at fixed time intervals without an observer being present. The installation in Cuba of such an advanced instrument for its time constituted a significant scientific advance.

This device was created through the collaboration of two men. Angelo Secchi, a Jesuit priest like Viñes, was responsible for its design, and a Frenchman, M. Salleron, manufactured it. The original model, constructed by Salleron at the Agriculture College at Grignon, France, did not include a thermometer, which was part of subsequent models such as the one Viñes used.

Fr. Angelo Secchi was born in Reggio, Italy on June 20, 1818. He joined the Jesuits at age 13 and worked over the years in many universities run by that order, among them Georgetown in Washington, D.C., where he taught physics and astronomy. Particularly known for his work in solar astronomy, he eventually became director of the Vatican Observatory. Secchi died in Rome on February 26, 1878.

Some of the records produced by the Secchi meteograph can still be found at the archives of the Institute of Meteorology. They are sheets 56 x 32 cm, ruled in grids of 5 x 5 mm. Data are recorded on these sheets from left to right as follows: temperature, atmospheric pressure, direction and speed of the wind, and amount of rain received. Each sheet contains the records for a period of seven days. It is interesting to note that these sheets are annotated on the margins in Viñes' own handwriting, with remarks about unusual readings or changes in trends.

The meteorological observations made during the period discussed in this book can be divided into two periods: those before and after the arrival of the Secchi meteograph. Starting in 1862, observations were made every two hours between 4:00 and 22:00 local time, although the 4:00 observations were suspended in 1863 and only restarted by Viñes in 1871, bringing the number of daily observations back to ten, a schedule that was maintained until 1903.

Records of these observations have been preserved in the Observatory, the product of sustained efforts made during those years.

The second period began three years after Viñes' arrival, with more complete records available after the installation of the meteograph, which provided 24-hour coverage to supplement the ten routine daily observations already in place. The better quality data provided by these direct observations allowed for the calibration of the meteograph data.

The ten daily observations included the barometric pressure, evaporation, rainfall, temperature, water vapor pressure, relative humidity, and wind direction and speed. In addition, cloud observations were made, as well as observations of the horizontal and vertical components of the magnetic field. All of these would later be supplemented by data obtained with equipment acquired during a subsequent trip Viñes made to Europe, as discussed later.

After the 1873 storm and during all of 1874, western Cuba did not experience any tropical cyclones (a tropical storm may have affected far eastern Cuba in July of that year). However, the next year a hurricane of large diameter and moderate intensity passed through a large portion of the island, starting with the former province of Oriente on the night of September 12, 1875, and moving out to the Gulf of Mexico a short distance east of Havana (Fig. 3). Damage was considerable throughout the whole coun-

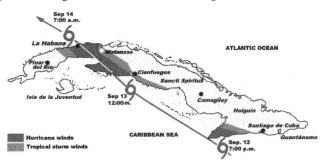

FIGURE 3: Areas affected by hurricane and tropical storm winds during the September 1875 hurricane.

try. This was the first hurricane whose forecast by Viñes is well documented in Havana newspapers. An extra edition of *La Voz de Cuba*, dated September 11, 1875, and urgently circulated in the streets of Havana, contained the following lines:

We have received at this late hour the following information with which Father Benito Viñes of the Company of Jesus, Director of the Observatory of Colegio de Belén, honors us, and which we feel compelled to disseminate to the public with the utmost urgency:

"...According to our sources, the Navy Headquarters has received telegraphs from St. Thomas and Puerto Rico...announcing that a hurricane had originated on the 8th in the Windward Islands.

I believe that this hurricane is likely to advance toward the NW and is likely to pass to the north and east of us at a considerable distance. Around the 13th at the latest we might feel its slight influence, if only the upper, cirrus parts accompanied by a movement of the barometer.

It would be good to alert the captains of ships intending to travel toward the North and East. As far as the travel to the W and in the Gulf, I do not believe that there is immediate danger.

These are only my rough estimates based only on the general laws of gyrating storms and a few short years of my direct observations." (La Voz de Cuba, 1875)

Reading this communiqué, especially the last few lines, one can conclude that Viñes had not issued forecasts before. He uses phrases that reveal his uncertainty, such as *"These are only my rough estimates" and "a few short years of my direct observations."* He did not use this type of expression again in subsequent press releases, after he became a seasoned forecaster.

This first forecast verified reasonably well, and thus resulted in Viñes' first success in the eyes of fellow scientists and the public at large. Witness the note published on September 15, in a Cienfuegos newspaper, which included comments on the forecast.

> *...Starting at 3 PM, the wind began to switch to the W, becoming ever stronger and accompanied by heavy downpours. By 5, the wind had begun coming from the third quadrant...*
>
> *...From the notes we have just communicated to our readers, it is clear that this phenomenon has only slightly affected the western end of the island...before disappearing towards higher latitudes.*
>
> *We are pleased to note that the predictions of the Belén Observatory have proven correct, as we had assumed yesterday.* (El Avisador Comercial 1875)

One year later, in 1876, Viñes forecast two other hurricanes with greater precision (Fig. 4). The first took place between September 15 and 17, entering the island near its eastern tip, and after affecting Guantánamo, went offshore heading west, and then northwest, crossing the island again near the city of Sancti Spiritus in central Cuba before reaching the Florida Straits.

FIGURE 4. Tracks of September and October 1876 hurricanes, showing areas of hurricane and tropical storm winds.

The following report on the September 1876 hurricane appeared in the press:

> ...*we have received the following press release from the wise meteorologist Rev. Father Viñes:*
>
> "*The telegrams received last night regarding the progress of the hurricane are now more precise and alarming. A hurricane that passed south of Puerto Rico is now SE of Cuba... [it] now poses a real and probable threat to us...*
>
> ...*the hurricane could well be near us by tomorrow... it is likely that by this afternoon we will have the first clear indications of the approach of the hurricane.* (Diario de la Marina 1876)

As had already occurred with the 1875 hurricane, the harbormaster was the first notified, and through him anyone preparing to sail out of Havana Harbor received timely warnings about the approach of the storm.

In October 1876, another hurricane, following a path typical for that month of the year, crossed the island in a northwesterly trajectory from the western end of the marshes of the Zapata peninsula (Ciénaga de Zapata) on the south coast, southeast of Havana, to a point on the north coast immediately east of the capital. Viñes noted the first signs of the approaching hurricane on Sunday October 15, almost 72 hours before the hurricane reached the vicinity of Havana.

On Tuesday, October 17, the *Diario de la Marina* published the following press release written by the meteorologist-priest:

Publisher of the Diario de la Marina

My Dear Sir.

From Sunday on the weather has taken an ominous turn. Starting yesterday, the 16th, there were indications of a rotating storm to the SE of us.

The drop in barometric pressure, the cirrostratus and cirrus uncinus, always precursors of the storm, the cirriform sky with solar haloes, the sunsets with a characteristic coppery hue of red, the low clouds moving with great speed from the NE, and the strengthening wind accompanied by alternating light and heavy showers constitute a sequence of events typical of rotating storms that no longer leaves us any doubt of its impending appearance...

...The slow pace with which these events are unfolding is proof that the vortex is approaching with very low velocity, which happens...as they turn in their path...so one can say that it is now turning, or soon will be.

As events unfold and data is gathered, I will keep you apprised of the different phases this phenomenon will go through, to the best of my understanding of it.

The following day, October 18, *La Voz de Cuba* published the following:

Editor of La Voz de Cuba,

Dear Sir:

...The vortex of the hurricane has entered the island near Trinidad. In all probability it is initiating its turning, and that is a very ominous development, since as its path curves it slows down and remains nearly stationary...

…The entire portion of the island bounded by a line be-tween Sagua and Trinidad (in central Cuba, to the east) and Havana (to the west) is currently experiencing the worst of the winds. Should the wind direction shift tonight to the E or ESE, we could experience even stronger winds.

Observatory of the Royal College of Belén, in Havana, the 18th of October 1876 at 2PM

Your humble servant,
Benito Viñes, S.J.

This strong and slow-moving storm produced a lot of damage in all areas affected by hurricane-force winds. In Havana, according to Gutiérrez-Lanza (1934), there was a three-hour period during the passage of the storm when the winds were calm.

On October 19, as the hurricane neared Havana, *La Voz de Cuba* printed an extra edition several hours ahead of its afternoon edition. Here are some highlights of Viñes' note:

We are now very close to the vortex of the hurricane, which I believe has passed near Alacranes and Güines (SE of Havana). It is urgent to prepare for every eventuality…

If winds from the NE and ENE persist, we will experience the calm of the vortex, and it is imperative to prepare for the wind to shift to NW. The wind shift will occur suddenly and with a terrible force. Shut the doors and windows facing SW. Ships should be well moored from that direction. Do not be lulled by the calm period. The winds that follow will be very strong and gusty…

The hurricane path is curving and it is moving very slowly…we just lost our best anemometer in one of the strongest gusts and only have that of the meteograph left, which has survived intact.

Yours...
Benito Viñes S.J.
Observatory of the Royal College of Belén, Havana,
19th of October 1876

P.S. While copies of this announcement were being written
the wind has become calm. This is the calm of the vortex.

The newspaper added these remarks to the communiqué above:

The hypotheses regarding the trajectory and speed of the hur-
ricane that the wise meteorologist made, based on his repeated
observations, have come to pass. This makes us presume that
the ones he is now making will also prove valid. Consequently,
it is vital that we take the necessary precautions, as an ounce
of prevention is worth a pound of cure...

The October 1876 hurricane became the "best forecast" of those
that had affected Cuba until that time. By that time Viñes could
count on having more practical experience and sharper skills in
detecting the signs of an approaching hurricane. In the hours prior
to, and during the storm, he increased the weather observations to
four per hour, no doubt with the intention of better sampling the
evolution of the weather during the passage of a hurricane. The
October 19 newspaper article continued:

Reverend Father Viñes, Director [sic] of the Royal College of
Belén has promised to send us a note summarizing the com-
plete set of observations made in said Observatory, which were
made every fifteen minutes for a period of 48 hours.

All the newspapers of Havana and the provinces praised the
accuracy of Viñes' forecasts. Phrases like "as was predicted by Fr.

Viñes" or "as Fr. Viñes envisioned" were common in these reports. Looking at the text of his press releases in 1876, it is interesting to see how Viñes appears to have gained in confidence in these forecasts, compared to his first, tentative ones.

After the passage of the first of the two 1876 hurricanes that hit Cuba, on September 16, *La Voz de Cuba* published an article entitled: "We are now free of danger" that expressed the relief felt by the citizens of Havana after the hurricane had moved out to sea. In that article, the editors began to pave the ground for public support of a novel idea: the creation of a Caribbean-based hurricane warning system headquartered in Havana. This is how Viñes, quoted in this article, broached this important subject:

> *The trajectory of a hurricane can be easily deduced from a few accurate observations…along various points near the path of the hurricane…*

Evidently Viñes realized that a network of telegraph stations set up to disseminate meteorological observations, especially in the eastern portion of the Caribbean, would give an exceptional boost to hurricane preparedness. This was a fortuitous moment for launching this idea, since the government authorities that could facilitate the establishment of such a network were dealing with the aftermath of these hurricanes, and Viñes was aware of the importance of favorable press for such an undertaking.

In fact, the various newspapers took up this idea with enthusiasm. Twelve days later, on September 28, *La Voz de Cuba* published an article entitled "A useful project" which ran four columns of a letter from Fr. Viñes:

> *…I dare propose a project, as simple as possible, with the purpose of establishing a network of simultaneous meteorological observations at set times and with the maximum feasible number of stations, transmitted at once by telegraph and later*

by mail to His Excellency the General Commandant of the Navy...with the goal of ameliorating the lack of observations and providing timely warnings of impending danger...The Matanzas tragedy is still fresh in our memory...

We know fairly well the laws that govern the course of these storms, but their origins are still unknown, and that may be because of the lack of a sufficient network of accurate observations in our region.

Further on in this article under the subheading *"Against the Hurricane"* Viñes was quoted giving further details of the proposed project:

"The plan we are proposing is also simple and doable, and all can contribute to it. It consists, as I've mentioned, of making a few standard observations at set times and only when the first signs of a potential hurricane are detected, or received by telegraph. Upon the given signal, the observations are to be transmitted to the General Command of the Navy, and finally to issue a warning from Havana to those points considered most at risk... (La Voz de Cuba 1876)

This plan, unfortunately, did not become "official" until ten years later, and even though Viñes always relied on information that the Navy Headquarters forwarded to him from ships or Caribbean ports, that network never functioned as had been hoped for. Later we will examine more details of its organizational plan. In retrospect, what stands out is the fact that this man was the first to envision ways in which the harmful impacts of hurricanes, these powerful and heretofore mysterious phenomena, could be mitigated.

As a driven man brimming with ideas that were very advanced for the late nineteenth century, Viñes realized that relying only on sketchy reports received from other locations made it impossible to get a clear idea of the structure and characteristics of hurricanes.

He soon hatched a plan to conduct an exploratory trip to inspect and investigate in situ all aspects of the hurricanes that had crossed the island in the previous 13 months. The observational data gathered in Havana could only give him a vague idea of events that had happened at various locations as witnessed and described by people from their individual perspectives, all connected to the same meteorological events.

So Benito Viñes proposed to set out from the Observatory and investigate firsthand the impacts the 1876 hurricanes had had throughout the island. To that end he sought the permission of Father Angel Gallo, Director of the College, to whom he detailed the objectives of his planned trip. No sooner had the last rains and the winds from the SSW from the last hurricane subsided than Viñes began planning his trip, to take place during the last days of October 1876.

Viñes knew that his chances of obtaining financial support for his inspection trip were at their highest in the aftermath of the hurricanes. He was by then regarded as a top scientist, professor of the College and Director of its Observatory, and well known to a public that was grateful for his hurricane forecasts. So it is no surprise to know that among the first to lend his support was the Captain General of the island, the top military and civilian authority of Cuba, who had been eager to meet him. As Distinguished Fellow of the Royal Society he appropriately sought support for his trip from that institution. Writing in his distinctive style, on folios adorned with the seal of the College, Viñes' proposal to the Academy reads as follows:

> *The hurricane of the 18th and 19th of October, in its slow and destructive march through the most populated and wealthiest section of this precious, but now ravaged island, devastating its productive countryside, and sowing ruin and desolation far and wide, was without doubt one of the most extraordinary events experienced here, and will take its place in the an-*

nals of meteorology and leave permanent scars in the memory of the people. It is for this reason that this phenomenon is worthy of study.

The wide extent of its colossal spires, the surprising magnitude of its circle of calm, its trajectory and the fact that its path curved in the vicinity of the island, its slow and halting trajectory and consequently the duration of its passage, the fact that its vortex passed over the most populated and richest part of the island, including the city of Havana itself, the destruction caused by the battering of the wind over such a wide area, the horrendous flooding that ensued and their accompanying landslides; these are reasons that justify, and also facilitate, the careful study of this phenomenon in all its details and phases of development.

I therefore have deemed it convenient, if not imperative, to take temporary leave from my duties at the Observatory in order to as soon as possible undertake a scientific inspection trip to various parts of the island, for the purpose of documenting the path of the storm and the trail of sorrows it has left behind in the memories of innumerable witnesses, gathering along the way data that will help me publish a memoir of the hurricane. These data, observed through my own eyes in the places where the disaster took place, and later sorted and analyzed, will be, I expect, of immense value, and will allow a rigorous scientific publication based on them.

God willing, my departure for this trip will be in a few short days, and I am fulfilling my duty as a member of the Academy you so capably lead to inform you and the Academy members of my plan so that, if you deem it useful, I can serve its interests in any way during my planned trip. I also humbly ask for your help in providing me with as many introductions to corresponding members in the provinces, or other persons that you deem could supply me with data and other help for this enterprise.

May God keep you.
Havana, November 3 1876
Benito Viñes, S. J.
Director of the Observatory
(ACC Archives)

This document was reproduced in part in the monograph entitled *"Notes Regarding the Observatory of Belén,"* although here we have reproduced the original complete version from the ACC Archives. Whether notified by means of this letter or as a result of previous private discussions, the Academy sent the College the following note in response, dated November 9:

Rev. Fr. Benito Viñes S. J.
Director of the Colegio de Belén Observatory

I have the honor of sending you the attached letter of intro-duction in which, in the name of this Academy, we request all men of science, and in particular the correspondents of this Institution, to lend you any manner of help in accomplishing the noble mission you have undertaken, and for whose success we offer our heartfelt wishes.

May God keep you.
Havana, November 4, 1876

Accompanying this note was a letter of introduction and recommendation from the Academy:

The Royal Academy of Medical, Physical, and Natural Sci-ences offers warm greetings to all men of science who might read this missive, and in particular to the corresponding mem-bers of this Academy. We have the honor of presenting to you the bearer of this letter, the illustrious Rev. Fr. Benito Viñes,

Distinguished Fellow of this Academy. We are most grateful for any assistance you are able to offer him in gathering information about the damages caused in your area by the hurricane, floods and landslides of last October, and thereby help him attain the worthy scientific and humanitarian goals of his research.

In the name of the Academy, best wishes for good health to all men of science.
Havana, November 4, 1876

In addition to these documents, there were recommendation letters written by influential friends, that introduced Viñes to any manner of people that might offer him help, lodging, and transportation along the way, including sugar mill owners, engineers, railroad administrators, and ordinary citizens.

Viñes began his journey almost immediately, since as he himself stated, it would be desirable to reach as many places affected by the hurricanes before the floodwaters receded to their normal levels and other indicators of hurricane damage were erased by the passage of time. The urgency for his early departure was to some extent heightened by the many reports of death and destruction that had reached Havana, some of which might have been exaggerated in their retelling. Since Viñes does not give the exact departure and return dates in the accounts of his travels, to the best of our knowledge he must have started his trip in the first half of November 1876.

In his monograph "Notes Concerning the Hurricanes of the Antilles" he writes:

Upon starting my journey, in light of the alarming, multiple and occasionally absurd accounts...of these events, my first and main objective was to visit the locations most affected by the flooding before the flood waters had completely receded...

[and] they were receding by the beginning of November, though very slowly. (Viñes 1877)

In addition he states in passing *"I was also planning to inspect as soon and thoroughly as possible the areas on the right side of the path of the hurricane of October 1876, with the aim of visiting them again on my return trip. I accomplished both purposes in the short time that was available to me."*

From this we can infer that in addition to the publically stated purposes of his trip, Viñes intended to compare his observational data against the damage that occurred in the right semicircle of the October 1876 hurricane, especially since Havana, where he made his observations, remained on the left side of the hurricane path. He was interested in comparing the damage on both sides along several traverses to obtain ground truth on the changes of wind direction and their subsequent effects.

Before concluding this narration, I should caution the reader that in all my explorations, my primary aim was to complete them to the extent possible before my departure from each location, and consequently in more than a few places I was unable to visit locations of potentially equal relevance to those I have cited, and I skipped others because of their lack of solid data and also because, as might be supposed, I added new locations to my visitation list as a result of new information that emerged during my inspection of other places.

Everywhere he visited, Viñes used a sophisticated and well-organized methodology. He recorded data in relation to the direction toward which trees had fallen, and if one tree was found on top of another, he used that information to determine the order in which they fell and thus the shift in wind direction that had occurred during the passage of the hurricane. Viñes estimated the peak level of floodwaters by taking the local topography into account. He interviewed eyewitnesses, pinning down the exact date and hour when events took place, whenever possible. No detail was overlooked or ignored. We shall later see how crucial these observations were to his subsequent work.

He found the fronds fallen from palm trees to be useful indicators of the wind direction, helping him pinpoint the path of the "vortex."

With respect to the evidence for the cyclonic rotation of the winds he says:

By sheer luck, in the short space of two leguas[11] one can see clearly marked the damage caused by the winds circulating around the small area of calm. Seeing such clear evidence of the swirling wind over such short distance, my travelling companions were filled with amazement upon contemplating what would seem an inexplicable phenomenon. Presented with such compelling evidence, we concluded that this type of phenomenon must have happened on multiple occasions in the past and should leave indelible evidence in place for long periods, without which the nature of cyclonic winds and gyrating storms would not have been understood. (Viñes 1877)

Viñes had to use other data sources on other occasions, since by the time he got to visit the affected areas, much needed repair work had begun erasing some of the signs left by the hurricane. Of the methods he used in these circumstances he says the following:

Luckily, in addition to data obtained in situ, I have endeavored to interview persons that weathered the storm at differ-

---

[11] A legua (Spanish for "league") was a widely used unit of length equivalent to 2.597 miles.

ent locations along its path, and even though some might not have had accurate memories of the event, or in some cases provided contradictory evidence, it was almost always possible to establish some facts that would allow me to form a considered opinion of the sequence of events witnessed. Some people also kept notes of the event, and we were often able to resolve questions that emerged during the course of the interviews by comparing them against the evidence on the ground. It is through the combination of all these methods that I was able to gather more or less complete data on the event.

Occasionally, where I least expected it, I was able to obtain valuable barometric and sometimes even pluviometric data.

In these investigations Viñes acknowledges having received the help of many people to whom he had been recommended by the Royal Academy, the colonial government, and sugar mill owners residing in Havana.

Ship captains were particularly helpful in this regard, attuned as they are to the weather. Also helpful were railroad managers and other educated people, but Viñes did not underestimate the accounts of simple county folk. He did not balk at changing transportation modes as needed, going from train to horseback, or horse-drawn carriage to travel on foot. He faced occasional danger travelling by small boat to places that were still flooded, and was accompanied by local guides to places that were remote and nearly inaccessible.

Something else from the trip left a profound impression on Viñes, more lasting than the unending torment of mosquito bites he endured: his direct witness of the suffering of the common folk that had lost all but their lives in the flood waters. This he recorded along with his scientific data, helping him connect the physical phenomena of hurricanes to the devastating effects they had on society.

In conclusion, if these and other no less exciting adventures of this memorable trip produced some merriment among our group, the misery we witnessed among the people affected left a deep impression in our hearts. We witnessed the tragic scenes of good and honest families driven to despair after having been forced out of their homes by the rising waters, and now taking refuge in improvised shacks or tents, some for the second time in a short period, in a few cases after moving their temporary shelters after the first storm's flooding to what they thought was safe, high ground, only to be made homeless again by the second hurricane's flooding. (Viñes 1877, p.10)

It is easy for those that know the Cuban countryside to recognize the types of humble dwellings that those living in remote areas call "vara en tierra," an improvised structure that can serve as shelter while the more comfortable traditional dwelling, the "bohío," is built with thatched roof and lumber walls, all with material taken from the ubiquitous royal palm. The traditional hospitality and courtesy of the men and women of the countryside are duly acknowledged by Viñes, who writes:

These good and honest friends, who will not be offended if I call them such, provided us with all manner of support and comfort, even as they were in the midst of such privations and misfortunes themselves. This was all the more welcome in light of our circumstances, which were often precarious too. (Viñes 1877, p. 11)

Viñes is parsimonious in recounting some details of his travels and only lists general locations. But based on his mention of certain reference points and checking against topographical maps of the area, it is possible to do a reasonable reconstruction of the places he visited. As far as place names, some deductions need to be made due to changes that have occurred over time, which made some names change their spelling and pronunciation or be renamed altogether. Some settlements have also been abandoned.

According to bibliographic sources, Viñes took four trips between the end of 1876 and the first half of 1877. The first three encompassed the western side of the country: the eastern side of Pinar del Rio province, Havana province, and western and central Matanzas province, as those provinces were defined in 1875. The fourth trip, of longer duration, took him to eastern Cuba, Santo Domingo, and Puerto Rico.

In planning these voyages, Viñes took into account reports he had read in the newspapers, and especially those related to hurricanes or tropical storms from ship reports received at the colonial Navy Headquarters and forwarded to him, whether those reports originated at sea or while the ships were in port.

Among the newspapers found in the Observatory's collections, compiled under Viñes' direction and used as a research tool by him, were *La Voz del Guaso of Guantánamo, La Voz del Comercio of Sancti Spiritus, the Diario de Cienfuegos, El Progreso of Cárdenas, El Eco Español* of Trinidad, *El Boletín Mercantil* of Puerto Rico, and, of course, the newspapers of Havana. Viñes used these sources to prepare a synoptic analysis of the hurricanes from the published narratives, and came to preliminary conclusions that he used to plan his trip so that he could test them with data gathered in situ.

In the first trip, according to his brief description, he left Havana for the city of Matanzas by railroad (Viñes 1877, 18-19). Once there, he made a short side trip to the nearby Monserrate heights in order to get an overview of the area. From Matanzas he went farther east to Cárdenas and thence to Colón to the south.

While in this general area, he visited places in central Matanzas province that were still flooded, located between the sugar mill formerly called "Santa Rita" (now René Fraga) and San Antón de la Anegada to the north. In this area, rowboats were frequently needed for transportation. The fact that large areas were still flooded during his visit and the date of the letter of recommendation from the Royal Academy place this visit as sometime in November 1876.

Once this segment of the trip concluded, he proceeded by the Central Railroad to various places in the vicinity of "el Roque," Jovellanos, Bolondrón, and Unión de Reyes, all in Matanzas province. From the latter place he went on horseback to a place named "El Galeón," adjacent to Zapata swamp, then entering deep into the swamp to visit a farm owned by a Mr. Crespo.

Turning once again toward the north, he continued through Sabanilla (now Gualberto Gomez) and the areas surrounding Alacranes, as well as Nueva Paz and Güines in the province of Havana, this time by train. From Güines he went south again toward the "Providencia" sugar mill, now called "Osvaldo Sánchez," and from there on horseback to the south coast of Havana province at Rosario beach.

Viñes would surely have made observations of the effects of a possible storm surge on the south shore of Havana province, but nothing about this appears in his writings, except for what he describes in his book as a "rising sea" (mar de leva), a term then used for this phenomenon. This omission was probably because that section of the south shore of the island did not experience a storm surge at all, as the winds in the area blew in succession from the north, northwest, west, and west-southwest, during the passage of the October 1876 hurricane, always blowing from land to sea without an onshore component.

Retracing his steps, Viñes went back to Güines to catch a train to Havana, stopping along the way for more observations at the "Loma de Candela" (Fire Hill), on the same railroad line. He was back in Havana toward the end of November or early December (Fig. 5).

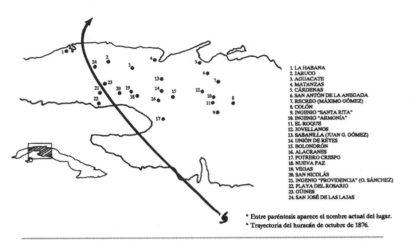

FIGURE 5. Places visited by Viñes in his first post-hurricane inspection tour east of Havana.

After resting at home for a few days, Viñes went on two more brief tours of western Havana and eastern Pinar del Rio Provinces near the city of Havana, where he managed to gather very useful data about the destruction caused by the hurricanes in that area (Fig. 6). He summarized both trips as follows:

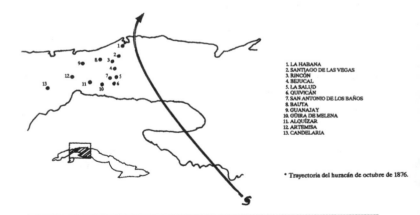

FIGURE 6. Places visited by Viñes in his second and third post-hurricane inspection tour west of Havana.

*A short while later I did two short excursions to Vuelta Abajo[12]
where I was able to gather numerous precise data points in the
vicinity of Guanajay, Alquizar, Quivicán, Artemisa, Las Man-
gas, Candelaria and various sugar mills, coffee plantations,
horse ranches, and farms of the region.* (Viñes 1877, p. 19)

These last two trips likely occurred during December 1876–
January 1877, perhaps with a short break home during the Christmas holidays. Adding to the speculation as to the timing of these
trips are his references of strong "nortes" during this period, a
feature of the winter climate of western Cuba.[13]

Once 1877 was underway, Viñes took his first trip abroad to
other Caribbean lands: the Dominican Republic and Puerto Rico,
both also affected by the hurricanes of 1876 (Fig. 7). As recounted
by him, this latter trip was longer than the previous ones, as it took
approximately a month and a half to complete. Although no exact
dates are available, I surmise that this journey took place between
February or March of 1877 to May of the same year at the latest.
We arrived at these dates based on the time it must have taken for
Viñes to edit his notes and prepare them for publication by the
Royal Academy for its 1877 publication date.

Viñes began this journey by train to Cienfuegos on the south
coast of the island where he boarded an eastbound ship making
stops in all the ports along the south coast of Cuba. He writes that
he was able to make useful observations during his stopovers in
Casilda, the port of the city of Trinidad, Tunas the Zaza, the port

---

[12] Vuelta Abajo or "Down Turn" in Spanish is the part of Cuba west of
Havana, where the island takes a turn to the southwest, departing from its
general southeast–northwest orientation

[13] During the winter season, cold fronts from North America occasionally reach western Cuba and beyond. In the Havana area, they are characterized by strong north winds and rough seas, accompanied by air temperatures
occasionally reaching the low 50's (Fahrenheit)

of Sancti Spiritus, Santa Cruz del Sur,[14] Manzanillo, and "other stops along the coast" before arriving in Santiago de Cuba, which he simply identifies as "Cuba" as was the custom at that time. From Santiago the ship went to San Juan, Puerto Rico, with stops in Puerto Plata, the Dominican Republic, and the Puerto Rican cities of Mayaguez and Aguadilla. Once in Puerto Rico, he visited inland areas around Humacao and the eastern shore of the island facing the island of Vieques.

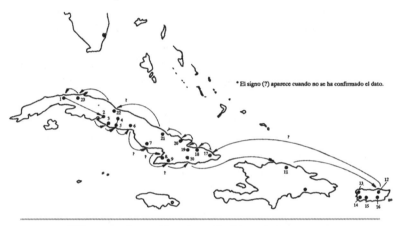

FIGURE 7. Viñes early 1877 post-hurricane Caribbean tour.

In Puerto Rico, Viñes applied the same methodology of carefully evaluating evidence of the hurricane damage in an to attempt to better define its path through that island, although by the time he reached that island much of the damage had already been repaired and he made note of the inevitably larger uncertainties resulting from this fact. Once back in San Juan he boarded a ship bound for Havana, this time making stops along the north coast of Cuba: Baracoa, Sagua de Tánamo, Mayarí, Gibara, and Nuevitas.

[14] The city of Santa Cruz del Sur, on the south coast of east-central Cuba was devastated by a hurricane storm surge on November 9, 1932, with a loss of 3,033 lives and many thousands injured or left homeless. It was the worst natural disaster Cuba experienced in the twentieth century.

Viñes left undone two projects he had planned for this trip: the first was an inspection of the coastline and hinterland in the vicinity of Guantanamo in far eastern Cuba. During that time, that area was a hotbed of the "Ten Year War" (1868–1878), the first large-scale armed insurrection of Cuban nationalists seeking independence for the island from Spain. The priest declined an offer by Spanish authorities to provide him with an armed escort for his travels through that part of Cuba, stating *"I did not deem it convenient to surround myself with armed people so that I could observe the aftermath of the hurricanes from close up..."* (Viñes 1877, p. 23). His use of the term "armed people" in this context was a not particularly flattering reference to soldiers of the Spanish Army. The second project Viñes had planned but was unable to accomplish was to visit the Isle of Pines (now called Isla de la Juventud), a large island south of Havana Province, from where he had received reports of widespread damage.

Another person may have accompanied Viñes for at least part of his travels abroad. References to that effect can be found in Fr. Gutiérrez-Lanza's Historical Notes on the Observatory of the College of Belén, and less precisely in Viñes' above referenced work. The following can be found first of those references:

> *The owners and managers of the various businesses he visited, and those of ships and railroads, not only provided him and his companion with free passage, but extended to him all manner of courtesies and made sure that others did too.* (Gutiérrez-Lanza 1904, p. 63, using as a reference Viñes' own published notes)

Viñes, on his part, limited his comments to the following phrase:

> *...not only did they give me and my companion free passage in their respective conveyances...* (Viñes 1877, p. 7)

Who accompanied Viñes in these travels? Based on these references we cannot be certain, but we can conjecture that the person in question may have been his friend Fr. Bonifacio Fernandez Valladares, an associate in the Observatory between 1875 and 1877 (Gutiérrez-Lanza 1904, p. 23). This priest, a professor of the College and a young man at the time, was later to lead a meteorological observatory at one of the Jesuit colleges in Spain, a fact that attests to the success of Viñes' mentorship of the young priest.

So it is not at all improbable that Fr. Valladares was the right person to assist Viñes in his travels. His help was probably made necessary not only in the face of the multiple challenges of travelling through a devastated landscape, but by the need to record data with sufficient precision.

The geographical reach of Viñes' four inspection tours during this period are summarized in the table below, which we assembled from data published in his *Apuntes Relativos* publication of 1877:

TABLE 1. Viñes' Hurricane Damage Inspection Visits of 1876–1877

Site visit	Probable dates	Region surveyed	Number of locations visited	Area surveyed (in km²)
First	November 1876	East Havana and Matanzas Province	23	11,700
Second	December 1876	Western Havana Province	12	2,500
Third		Eastern Pinar del Rio Province		
Fourth	January-March 1877	Cuba, Dominican Republic and Puerto Rico	29	8,800
TOTAL			64	23,000

The general criteria Viñes used in these trips to assess the extent of hurricane damage remain valid to this day, and these visits also have the historical distinction of being the first scientific expedition conducted in the Caribbean basin for a meteorological purpose, and more generally constitute the first scientific study of the impacts of natural disasters in the region as distinct from a merely descriptive chronicle of these phenomena.

Viñes' main accomplishments as a result of these trips can be briefly summarized as follows:

1) He was able to determine with precision the track of the hurricanes of 1875 and 1876 as they passed over the island of Cuba.

2) He either obtained for the first time or confirmed previous reports of the strength of these hurricanes as they travelled over land.

3) He established a cause-and-effect relationship between these natural phenomena and their impact on society and the environment.

4) He gathered empirical data that allowed him to test then commonly accepted hypotheses about the structure and dynamics of tropical cyclones.

5) He proposed the construction of a canal to provide better drainage to the area known as the "Plains of Colón" in order to prevent future flooding. That proposal was finally realized in the early decades of the twentieth century, when the Roque Canal was constructed in the province of Matanzas.

Notwithstanding the achievements listed above, these expeditions fell short of accomplishing some of the objectives Viñes had set for them. Among these shortcomings were the following:

1) He was unable to visit the Isle of Pines (now officially called Isle of Youth) south of Havana Province, where the October 1876 hurricane caused serious damage. We do not know the cause of this important omission.

2) Viñes was not able to visit several locations that were hit by the September 1876 hurricane in the eastern provinces of Cuba because the Spanish army was conducting military operations against insurgents in the area at the time. The possibility of being caught in the crossfire or being taken hostage made travel to the countryside too risky.

3) Viñes found scant instrumental meteorological data in his visit to Hispaniola, and was forced to use unreliable second-hand reports by the local press.

By the middle of 1877, once back from his inspection tours, Viñes presented his findings to the Royal Academy, as recorded in their Annals (Anales de la Real Academia 1877, v. 14, p. 230). This marked the culmination of more than seven years of careful observations of all manner of phenomena related to hurricanes. A few months later, these results were published in full in the book entitled "*Notes concerning the Antilles Hurricanes of 1875 and 1876*," which included the full text of all the oral presentations Viñes made on this subject.

"El Iris," a publishing firm located at 29 Obispo Street, printed this landmark book, which firmly established Viñes as a pioneer in the scientific studies of hurricanes. This work put forth new hypotheses, based on reliable observations of the behavior of recent hurricanes made from various points in the tropical Atlantic, Caribbean and Gulf of Mexico. The original manuscript can be found in the ACC Archives as part of the documentation on Fr. Viñes originally kept at the Observatory.

This work marks the conclusion of the first stage of Viñes' contributions to hurricane science, which culminated with his listing of general principles (which he called "laws") to be used in the forecasting of hurricanes, based in his years of experience on this matter, and which he put into practice with his successful forecasts of the hurricanes that affected Cuba on 1875 and 1876. His positive results put Viñes in a position to harbor loftier ambitions.

The Wreck of the Steamship Liberty

Another important event that, among others, firmly cemented Viñes' reputation as a meteorologist, was the sinking of a steamship whose captain chose to ignore Viñes' well-publicized forecast of a dangerous hurricane along his intended route.

Near midnight, during the night of September 13 to 14, 1876, the Navy headquarters at Havana received a telegram from San Juan, Puerto Rico, stating that a hurricane had passed just south of the city during the morning of the 13th, inflicting heavy damage to that island, along with winds of 100 kilometers per hour and a minimum pressure of 29.5 inches of mercury.

Approximately an hour later another telegram arrived, this time from Santiago de Cuba, informing that the islands of St. Kitts, St. Croix, and St. Thomas had experienced a storm during the night of the 12th. Several people were reported drowned in those islands, along with two missing ships. The telegram from Santiago went on to mention that Jamaica had begun experiencing stormy conditions on the 13th and that the weather in Santiago was deteriorating even as the telegram was being sent, with falling barometric pressure and strong gusty winds from the northeast.

Father Viñes, a few blocks away from the Navy headquarters, and eagerly awaiting further news about the hurricane, was no doubt immediately notified of these ominous developments. With these new data in hand, Viñes promptly sent an urgent update to the local press. Describing the newly received telegrams as containing "more precise and alarming" information, he mentioned in his dispatch that the second, or less likely, scenario (the "second hypothesis" in his words) that he had mentioned in a previous press release would likely be confirmed. That is, with the hurricane now tracking farther to the west than he had originally anticipated, it now posed a greater danger than Viñes had previously foreseen.

As Viñes was communicating the disturbing new forecast to the harbormaster and the press, several ships were preparing to sail from Havana harbor. Among these was the SS *Liberty*, whose crew, under the command of John T. Sundberg, was finishing preparations for a trip to New York. The *Liberty* was loading a cargo of sugar, palm leaf, and tobacco valued at $50,000.

Captain Sundberg, a fifty year old Swedish-born resident of Booklyn, was an experienced shipmaster who had been in the employ of the James E. Ward & Co. shipping line for three decades. He had previously commanded the sailing vessels *Topeka* and *Cardenas* and the steamship *Cuba* (formerly the *Fort Morgan*), all belonging to the same firm, and all involved in the Cuba trade.

The *Liberty* was a ship of substantial size and unusual strength. "Her length was 332 feet; breadth of beam, 34 feet 7 inches, and depth of hold 25 feet; her draught was 16 feet, and she could carry 1,229 tons. She had direct acting engines sixty-five inches by forty-four" according to the now defunct *New York Herald* (New York Herald, September 26, 1876). The *Liberty* was built in 1864 by the George Lynn shipbuilders of Philadelphia. In 1868-1869, it underwent a major upgrade in engines, equipment, and passenger accommodations, which included amenities such as new paint and carpeting, and even a new piano in the passenger lounge.

At Viñes' suggestion, the harbor had already been closed to eastbound traffic. Captain Sundberg, his ship loaded with valuable cargo, seeing no visible indications of deteriorating weather, with the ship's engines running and a tight schedule to follow, ignored the stern warnings of the harbormaster, and the *Liberty* sailed off for New York on schedule, later on September 14, in fair weather with no passengers on board, just as the dangerous hurricane Viñes had forecast was beginning to cross eastern Cuba on its way to the Straits of Florida.

Ten days later, on Sunday, September 24, an article appeared in the New York Herald with the headline: A MISSING STEAMSHIP. The article went on to state:

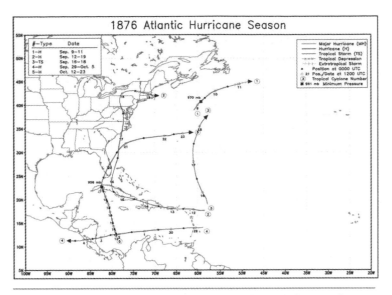

FIGURE 8. Track of 1876 hurricanes. The hurricane that sank the steamship Liberty is Hurricane Number 2. (Wikipedia/NOAA)

"*No little anxiety exists in this city with reference to the safety of the Havana steamship Liberty, which left that port ten days ago since which no tidings have been received of her... Last evening a Herald reporter called upon Mr. Ward, at his office at the foot of Wall Street when he made the following statement: We have no news of the Liberty since she left Havana. I do not think she is lost, but I think that she has become disabled. She has a wonderfully strong frame and can stand any amount of bad weather. You may remember the case of the Washington, of Messrs. Clark & Seaman, which sailed for New Orleans from here of which no news was received for upward of three weeks and then news was received from the Bahamas that she had reached a small island under sail, with damage to her propeller, so I cannot consent to take a gloomy view of the Liberty's fate for some time. I do not know how much cargo she has on board, but I think she is somewhat light. To* *the best of my belief she has no passengers on board…she is worth about $90,000 and is only partially insured, the risk being placed in this city."*

Two days later, the New York Herald gave the following update to its subscribers:

THE STEAMSHIP LIBERTY
FOUNDERED ON HER WAY FROM HAVANA TO NEW YORK
THE CREW SAVED

The Herald of last Sunday contained an announcement that the overdue steamship Liberty, *bound from Havana to this port, had possibly foundered in the recent gale. Yesterday afternoon, the supposition was verified, as the following dispatches will show. The first is from the late commandant of the unfortunate craft to the owners:*

Lewes, Del. Monday
Steamer Liberty foundered in hurricane of the 17th inst.
Crew all safe

Shortly after the receipt of the foregoing, the following telegram was received:

Lewes, Del. 5 PM
The steamer Liberty was abandoned on the 17th inst.[15] in latitude 33 north, longitude 75 west. The crew all taken on by the schooner Yellow Pine.

---

[15] Abbreviation often used in newspapers of the time meaning "of the current month."

In conversation with Mr. Ward, the owner, he said that Captain Sundberg was an excellent shipmaster, and doubtless did all that human ingenuity could suggest to save the vessel before he decided to abandon her. (New York Herald, September 26, 1876)

In an album of newspaper clippings on the maritime disasters of 1876 and 1877 held in the collections of the New York Public Library, there is a more detailed (and vivid) account of the fate of the Liberty. Unfortunately the publisher and publication date are not recorded in the album. This is what it states:

THE WRECKED STEAMER
A GALLANT RESCUE OF THE CAPTAIN AND CREW
BY THE AMERICAN SCHOONER YELLOW PINE

With regard to the foundering of the S.S. Liberty, belonging to James E. Ward & Co. of 113 Wall Street, while on her way from Havana to this port, her captain, T. P. Sundberg, makes the following report:

The vessel sailed from Havana on the 14th inst. with a cargo of sugar, palm leaf and tobacco, valued at $50,000. There were no passengers on board, and the crew consisted of twenty-five persons, all told. On the 16th, in the afternoon, a heavy gale from the southeast was experienced.

It increased in violence during the night, and blew a perfect hurricane toward morning. The hurricane continued during the day of the 17th, and at 2 PM in latitude 33.10, long 77.15, a heavy leak was discovered by the rapid filling of water in the hold.

Hopes were entertained of keeping the vessel afloat with the pumps until about 6:30 PM when the water reached the main furnaces and extinguished the fires.

The last ray of hope then fled from the minds of the crew, and they made up their minds for a watery grave as they saw the hold rapidly filling and no means of preventing it.

The life-boats were got ready for rise at short notice, but no one expected to save his life by means of them, because, though the wind had somewhat subsided by this time, the sea was so very rough that no boat could remain together an instant on it.

In the midst of their gloom and despair, which every one was sure his last hour had come, the joyful cry of "hail, ho!" was uttered by the lookout, and was greeted by three hearty cheers by all hands.

Signals of distress were hoisted, and a few hours after the whole crew were on board the schooner Yellow Pine, Captain Clark, bound for the Delaware breakwater.

There were, however, but provisions enough on the Yellow Pine to last her own crew for seven days, and some were obtained from the steamer by a few of the crew, who returned to her and placed lights on board to enable the Yellow Pine to remain close to her all night.

The intention was to save some of the cargo should she still be afloat the following morning, but when daylight broke she had disappeared. The crew all returned to this city to-day.

The return of the crew of the Liberty to New York on September 26, 1876 (which included the Captain's son, C.L. Sundberg, who was listed as a porter by the Herald) was announced by the *New York Tribune* and the *Brooklyn Daily Eagle*. Most of the crew members were residents of Brooklyn, which did not become a borough of New York city until 1896. One day later, the *New York Times* reported that one of the *Liberty* crewmembers had died as a result of exposure.

Dramatic as the wreck of the *Liberty* was, it was not, by far, the biggest loss of life or property that resulted from this particular hurricane of September 1876. The aforementioned album also contains a clipping from an unidentified newspaper about another shipping disaster, the wreck of the *Rebbeca Clyde*, reported on September 19. The *Rebbeca Clyde*, which was travelling from Wilmington, North Carolina to Baltimore, ran aground at Portsmouth, off Pamlico Sound, taking thirteen lives with it. *"Captain Childs, two mates, two engineers, three seamen, the steward, two coalheavers and one passenger named Whilton were reported lost... Much of the cargo has been washed up on the beach. The survivors speak with tearful earnestness of Captain Childs. He did everything that mortal man can do to save his ship and the lives of those who were intrusted [sic] to his care."*

The hurricane of September 1876 left a path of destruction up and down the eastern seaboard of the United States. Clippings from the album also mention reports from Baltimore of washouts delaying trains, and several schooners sunk or damaged by the storm, including one, the *D. E. Woeford*, which was headed for Richmond with a full cargo of coal. Its crew, according to the report *"climbed up to the mast, where they remained all night, and were rescued next morning."*

The *New York Times* September 18th edition gave a detailed account of the storm as it played out in the New York City area and nearby cities such as Philadelphia, Poughkeepsie, Milford, and Providence. Referring to the *"torrents of rain accompanied by a gale"* that New York City experienced, the paper mentioned that it was *"no doubt the remnants of the hurricane that visited the West Indies a short time ago. Its coming had been predicted for several days, and its course was easily marked by its steady progress from Key West along the Atlantic coast."*

Father Viñes' forecast, rapidly disseminated far and wide by telegraph, the then cutting-edge medium of communication, had undoubtedly provided the early warnings mentioned in that *New York Times* article.

As for Captain Sundberg, who died in his home at 937 Bedford Avenue in Brooklyn on January 4, 1903 at age 76, it is difficult to judge too harshly his decision to sail from Havana on that fateful day after being warned about the possibility of encountering a hurricane in his projected route to New York and ignoring the warnings of the harbormaster. Hurricane forecasting was in its infancy, after all.

But not a single resident of Havana boarded the *Liberty* on September 14, in spite of all the amenities it provided, thanks to the work of a very thin priest who had made such careful cloud observations from that small school located on Compostela Street, between Luz and Acosta Streets, in Old Havana.

FIGURE 9. Photo of the *Liberty* (U.S. National Archives).

Childhood, Youth, and Vocation

2

I have endeavored to work with the diligence required to discharge a sacred duty, and by virtue of my desire to understand physical phenomena at a deeper level, especially the unfolding of the terrible majesty of hurricanes. (Diario de la Marina 1875)

Benito Viñes was born at 11 AM on September 19, 1837, in the small village of Poboleda, deep in the mountains west of Tarragona, in Catalonia, Spain. According to the custom of the era, he should have been named after the saint whose feast day was celebrated on that date, but in his case that tradition was not followed. His father was named Carlos, who in turn was the son of Benito; his mother was María Teresa Martorell. The infant was baptized the following day, September 20, and given the full name of Carlos Benito José Viñes Martorell, although as an adult he was always known as Fr. Benito Viñes, or simply "Father Viñes."

This is how his baptismal record reads:

On the 20th of September of 1837, at the baptismal font of the Parish Church of St. Peter in the village of Poboleda, Archdio-

cese of Tarragona, the signatory below solemnly baptized Carlos Benito José, natural born son of Carlos Viñes and María
Teresa Martorell spouses from this village, who was born at 11
in the morning the day before. Paternal grandparents Benito
Viñes and Raimunda Viñes. Maternal grandparents Joseph
Martorell and María Casanovas. The godparents were Miguel
Martorell and Teresa Pahí y Viñes, whom I instructed on their
obligations as the child's spiritual family to indoctrinate the
child in the Faith.

José Salvador, Parish Priest (Archdiocese of Tarragona
Archives 2003)

His godparents were his maternal uncle Miguel Martorell and
his wife Teresa Pahí. The baptismal records can be found in the
registry of St. Peter's Church, as the parish church (built in the
mid-1700s) was then named, under entry number 62 of Folio 172,
volume 6. Benito had two siblings: José and Teresa, of whom little
is known. Benito's father died in 1846, when the youngster was
only 9 years old.

Poboleda, located on the banks of the Ciurana river about 12
kilometers from the town of Falset, had at the time a population
of fewer than 1,000 inhabitants who made a living growing grains,
potatoes, almonds, and olives and producing olive oil and wine.
Poboleda's origins date back to the economic activity sparked by a
nearby Cistercian monastery. Tarragona, in turn, dates back to Roman times and had formed part of the Kingdom of Aragón before
Spain became a united country in 1492. Gutiérrez-Lanza, in one of
his writings, recounts a conversation in which Viñes told him that
Don Carlos, his father, used to be consulted by villagers wanting
a forecast of the weather, as sometimes happens with respected
villagers in traditional rural communities.

As, perhaps, Poboleda's most distinguished native son, the
house where he was born is now identified with a historical marker.

FIGURE 10. Poboleda as it appears today. Photo by DAnRov.

FIGURE 11. House where Viñes was born. (Photo by the author)

Records of Benito's early education are sketchy. While in secondary education, between 1850 and 1855, he received grades of meritísimo, the highest possible, in Latin and rhetoric. On May 12, 1856, he entered a seminary of the Jesuit order to begin his advanced studies and preparation for the priesthood. At that point he was yet to turn 19.

In an 1893 article published in the *Diario de la Marina* we learn that he did his novitiate in Mallorca and received his scientific training at the University of Salamanca where he attained the rank of Catedrático in Physical and Natural Sciences. Although we have no independent verification of these facts, they are in keeping with other indirect references.

Viñes remained in Spain for the first 31 years of his life. The year 1868 was an important one in the history of Spain. Queen Isabel II was overthrown by a revolution headed by General Serrano. Two years later, Amadeo, Duke of Aosta and son of the King of Italy, was elected King of Spain. The 1868 revolution had a significant radical component and included the expulsion of Jesuits from Spain. As a consequence, Viñes and many of his Jesuit colleagues sought refuge in nearby France.

According to the second person to write a biography of Viñes, the Cuban Jesuit, Antonio López de Santa Anna,[16] Benito Viñes began his academic preparation for the priesthood in September 1850 and completed it five years later. Lopez asserts that Viñes' grades ranged between sobresaliente (Spanish for outstanding) and meritissimus (Latin for highest merit). By 1856 Viñes began his novitiate as a Jesuit in Mallorca, in the Balearic Islands, and later transferred to the Novitiate College of Loyola, also in Spain. Lopez notes that in the records of his life as novitiate, he was recognized for his aptitude for learning and his unblemished

---

[16] López, A. (1957): *Contribución a una biografía completa del padre Benito Viñes, S. I. Célebre meteorólogo de las Antillas,* Taller de Artes Gráficas de los Hermanos Bedia, Santander, 96 pp.

discipline, and his "relative frailty and scant physical strength" were attributed not to illness, but rather "to his innate physical constitution."[17]

López states that Viñes was in France in 1868-1869, living in one of the residences that had been hastily set up to house Jesuit refugees from Spain in the cities of Laval and Poyanne. It was in the latter city that he was likely ordained as a priest. According to López, on November 19, 1869, colonial government officials in Havana gave permission for a group of Jesuits to travel to Cuba, among them the newly ordained Fr. Viñes. The biographer states that the ship carrying the Jesuits left the French port of Saint Nazaire in February 1870, finally arriving in Havana on March 8 of that year.[18]

Viñes arrived in Cuba from Europe with a solid education, and his assignment to head the Belén Observatory undoubtedly resulted from the skill he had demonstrated in his studies of physics and the earth sciences. In addition to his native Spanish, he was fluent in Latin and French and could read English. Reading in these languages, he had become acquainted with the physical, meteorological, and astronomical literature of the time. He had also studied some geology, and, of course, the Roman classics. In his writings he made references to many pioneering works in the atmospheric sciences, which formed the theoretical framework of his initial research work in Cuba.

As it may be of interest to the reader, we list these references below. Some were undoubtedly read by Viñes while in Spain or France. Others, given their date of publication, must have been acquired in Cuba. Not all references are fully described in Viñes' notes—in some he mentions only the author's name—but it is possible to deduce the full information, based on the works that

---

[17] Loc. cit., 28.

[18] Loc. cit., 48.

had been published and were available to him at the time of his writing. Listed chronologically, these references are as follows:

Espy, James[19]	*The Philosophy of Storms.* Boston, 1841.
Redfield, William C.[20]	*On Three* [sic] *Several Hurricanes of the Atlantic and Their Relation to the Northers of Mexico and Central America.* New Haven, 1846.
Reid, William[21]	*The Law of Storms.* London, 1850.
FitzRoy, Robert[22]	*Livre de Temps* (Weather Book) Paris, 1850.

---

[19] James Pollard Espy (popularly known as the Storm King) (May 9 1785–January 24, 1860) was a U.S. meteorologist. Espy developed a convection theory of storms, explaining it in 1836 before the American Philosophical Society and in 1840 before the French Académie des Sciences and the British Royal Society. His theory was published in 1840 as *The Philosophy of Storms.*

[20] William Charles Redfield (March 26, 1789, Middletown, Connecticut –February 12, 1857 New York City) was one of the founders and the first President of the American Association for the Advancement of Science formed in 1848.

[21] Sir William Reid (1791–1858) was a British soldier, administrator, and meteorologist. Reid was sent to the Leeward Islands in 1831 to direct the task of reconstruction after the Great Barbados hurricane of 1831. During his two-and-a-half-year stay he became absorbed in trying to understand the nature of North Atlantic hurricanes, which led to a lifelong study of tropical storms. He published *An Attempt to Develop the Law of Storms by Means of Facts* (1838; third edition, 1850) and *The Progress of the Development of the Law of Storms and of the Variable Winds* (1849).

Kämtz, Ludwig Friedrich[23]   *Curso de Meteorología.*
                                Paris, 1858.

Piddington, Henry[24]   *Guide du Marin sur la Loi
                                des Tempetes.* (The sailor's horn-
                                book for the law of storms)
                                Paris, 1859.

José María Tuero[25]   *Tratado Elemental Aplicado
                                a la Náutica de los Huracanes.*
                                Madrid, 1860.

Poey, Andrés[26]   *Table Chronologique des Quatre
                                Cent Cyclones.*
                                Paris, 1862

---

[22] Vice-Admiral Robert Fitz Roy RN (5 July 1805–30 April 1865) is probably best known as the captain of the HMS *Beagle* during Charles Darwin's famous voyage. He was a pioneer in the emerging science of meteorology and authored the "Weather Book," published in 1863. In 1854, he was appointed chief of a new department to deal with the collection of weather data at sea, which eventually became the Meteorological Office of the United Kingdom.

[23] Ludwig Friedrich Kämtz (1801–1867) wrote an early, very influential book on meteorology that was translated into numerous languages. Kämtz conducted research on the characteristics of the daily cycle of temperature, air pressure and humidity.

[24] Henry Piddington (1797–1858) was a British official in colonial India and is credited with coining the word "cyclone" to describe rotating storms.

[25] Spanish naval officer and meteorologist.

[26] Andrés Poey (1826–1919) had a role in the founding of the Havana Observatory, and he served as director of an observatory in Mexico during the reign of Maximilian.

Marie-Davy, Hippolyte[27]	*Météorologie. Les mouvements de l'atmosphère et des mers considérés au point de vue de la prévision du temps.* Paris, 1866.
Bridet, M. (Hilaire Gabriel)[28]	*Ètude sur les Ouragans de l'Hémisphre Austral.* Paris, 1869.
Meldrum, Charles[29]	*Notes on Forms of Cyclones in the Southern Indies Ocean, and on some of the Rules for Avoiding their Centres.* London, 1873.
Wilson, W.G.[30]	*Report on the Midnapore and Bourdwan Cyclone of October 1874.* Calcutta, 1875.

---

[27] Edme Hippolyte Marié-Davy (1820–1893) was a French chemist and inventor. In the 1860s, he was Deputy Director of the Paris Observatory, in charge of meteorology He devoted himself to the study of local thunderstorms, following the destructive storm of November 14, 1854, in the Crimean War.

[28] Book available through Google Books.

[29] Charles Meldrum (1821–1901) founded the Meteorological Society of Mauritius and studied the climatology of tropical storms in the Indian Ocean, including their duration, the distance they travelled and the areas over which the wind blew with the force of a strong gale.

[30] W.G. Wilson, M.A., L.C.E, wrote this book while he served as Meteorological Reporter to the Government of Bengal.

Flammarion, Camille[31]   *L'Atmosphère.*
1873 French Edition and
1875 Spanish Edition.

Faye, Hervé[32]   *Défense de la Loi des Tempêtes.*
Bureau des Longitudes.
Paris, 1875.

Loomis, Elias[33]   *A Treatise on Meteorology.*
New York, 1868.
Contributions to Meteorology
American Journal of Sciences
and Arts
January–July 1877.

It is interesting to note that Viñes does not mention two works that were well known at the time: *The progress of the development of*

---

[31] Nicolas Camille Flammarion (1842–1925) was a French astronomer and author of more than fifty titles, including popular science works about astronomy and meteorology, some early science fiction novels, and several works about spiritism and related topics.

[32] Hervé Auguste Étienne Albans Faye (1814–1902) was a French astronomer. His discovery of "Faye's comet" won him the 1844 Lalande prize and a membership in the Academy of Sciences. He was as President of the Bureau of Longitudes, a scientific institution charged with synchronizing clocks around the world. He also served as Minister of Public Instruction during the French Third Republic.

[33] Elias Loomis (1811–1889) was a prolific author and professor at several American universities, including the City University of New York and Yale, he known for his research on the aurora borealis, especially the 1859 event, which was seen as far south as Cuba. He plotted weather maps to analyze two storms over the United States in 1849, using symbols to represent weather variables that became the prototype for what later became the "station model" used in weather maps today.

the law of storms, and of the variable winds: With the practical application of the subject to navigation; illustrated by charts and woodcuts by Sir William Reid (1849),[34] a book that had already been translated into Spanish; and a work on Philippine "baguios," the local name for typhoons, written in 1873 by the Spaniard Manuel Villavicencio.[35]

In addition to the works cited above, Viñes had at his disposal several periodicals, which the Observatory received either as a subscriber or by way of data exchanges with similar institutions. Among these we found the well-known American "Weather Maps" and various annual data summaries from astronomical and magnetic observatories, mostly from European countries. That constant stream of information helped shape Viñes' views, which had been primarily influenced by French meteorological thought, as was the case with other Cuban meteorologists of the time.

Viñes learned English after his arrival in Cuba, and could then read in their original language the works of authors such as Piddington or Fitz-Roy, which he had previously read in French translation. Undoubtedly, given the number of French scientific publishing houses of the time, Viñes had much easier access to the scientific literature in the well-equipped libraries of the French seminaries where he lived before arriving in Cuba.

Of course, Vines' library probably included a significant number of texts of philosophy, theology, and similar disciplines that formed part of his education as a Jesuit priest and must have played an important role in his worldview. On the other hand, we have no evidence of Viñes having read the important works

---

[34] This work contains further research on storms made by Reid after the publication of his *Attempt to Develop the Law of Storms* in 1838. (Text is available online at http://libraries.ucsd.edu/speccoll/weather/index.html.)

[35] Manuel Villavicencio's work is cited in another book on baguíos: *Baguíos o ciclones filipinos: estudio teórico-práctico,* by José Algué, S.J., published in Manila, 1897.

on hurricanes by Fernandez de Castro[36] or Desiderio Herrera,[37] which are classics in the field.

Shaped by the discipline of his training as both priest and scientist, Viñes was diligent and methodical in his work. Proof of this is his life-long habit of keeping notebooks where he recorded notes on a daily basis, invaluable documents that have, unfortunately, been lost. Gutiérrez-Lanza writes the following about Viñes' notebooks:

> *Every time thin cirrus clouds appeared over the skies of Havana, Fr. Viñes took note of their shape, position and the direction and speed of their movement. One or two days later a ship would arrive in port, or a telegram would arrive announcing that a cyclone had been observed on a particular date and location. Father Viñes would then consult his notebook and would find that the cirrus clouds emanated from the cyclone coordinates and their convergence, if any, pointed in the same general direction. This he found over and over again, hundreds of times.* (Gutiérrez-Lanza 1936, p. 20)

The observations recorded in his notebook served to supplement the press clippings of his published forecasts. This routine set a precedent for Viñes' successors at the helm of the Observatory, from Lorenzo Gangoiti to Rafael Goberna, who carried on this work for 87 years.

We have no direct evidence that Viñes taught classes at the College, even though Belén was a religious institution dedicated to teaching. Viñes was not, however, isolated in an ivory tower;

---

[36] Manuel Fernandez de Castro (1822–1895) wrote "Estudio sobre los huracanes ocurridos en la isla de Cuba durante el mes de octubre de 1870" (Study of the hurricanes that affected Cuba in October 1870).

[37] Desiderio Herrera (1792–1856) wrote "Memoria sobre los Huracanes en la Isla de Cuba," published in Havana in 1847.

he shared his knowledge and skills freely with his colleagues and students and his work served as model of the scientific method. He undoubtedly taught science classes while at Belén.

Viñes trained several of Belén College's priests and religious brothers in the application of the scientific method so they could serve as weather observers and help him maintain the Observatory's instruments. Among the first to work with him was Father Tomás Ipiña, who worked with Viñes in 1873 and 1874 and went on to become rector of the College in 1881.

Noteworthy among these disciples was Father Bonifacio Fernandez Valladares, his assistant between 1874 and 1877, who, as already mentioned, accompanied Viñes in his travels of 1876 and early 1877. Father Valladares was assigned elsewhere by his superiors in 1877, but returned to Belén in 1882-1885 before being appointed director of the Observatory of the Jesuit College at Oña, located in Burgos, Spain. Gutiérrez-Lanza quotes Valladares' recollections of his time with Viñes in Havana:

> "Of reverend Father Viñes I can attest as a witness of the years I worked alongside him, that the observations of cloud motion played no small part in his skillful forecasts... Around 6 PM, other, lower clouds appeared to the north, and he took prompt note of their direction of movement." (Gutiérrez-Lanza 1904, p. 61)

Another Viñes collaborator was Pedro Osoro, who worked closely with him between 1881 and 1883. Osoro had previously worked at the Belén Observatory between 1866 and 1870, before Viñes' arrival in Cuba. In addition to Osoro, father Mauricio Cid worked as an adjunct in the Observatory between 1877 and 1880.

Of Viñes' collaborators, Brother José Alberdi stands out. He began assisting Viñes in 1885 and served under him until Viñes' death. Alberdi worked in the various tasks of running the

Observatory: reading instruments, communicating with the press, and preparing various publications produced by the institution. Brother Alberdi's assistance was invaluable; one can see numerous notes in his handwriting accompanying press clippings in the archives of the Observatory.

The Observatory's routine tasks during Viñes' time were also shared by seminarians studying for the priesthood at the College. The last of these to help Viñes was Juan Gilibert, who worked at the Observatory between 1891 and 1897, four years after Viñes' death. His main task was to prepare the typesetting of data tables for publication, a labor that had become too challenging for Viñes with his declining health.

Last but not least in this list of collaborators is Father Mariano Gutiérrez-Lanza, who arrived in Cuba in 1891, shared a close friendship with Viñes for two years, and took over as director of the Observatory years later. Gutiérrez-Lanza wrote his memoirs, both in his capacity as historian of the institution and as biographer of his most famous predecessor.

Not only was Viñes a pioneer in the study of tropical meteorology in Cuba, but he was also one of the first to train others in this discipline and to educate the general public and serve the community by issuing forecasts, using the printed press, the most effective medium of communication of the time.

Evidence shows Viñes received as good a scientific training as possible for a man of his time and circumstances. In his training as a scientist he was influenced, more or less directly, by the major scientists of the time, both in Europe and in Cuba, in the latter case by the forum provided by the Royal Academy of Sciences of Havana. As an active participant in that institution, he was able to share ideas with the best and the brightest men of science in the island. Dr. Arístides Mestre, an Academy member, wrote in a Viñes obituary "the proud Academy...received on many occasions the benefits of that tireless and vast intellect" (Annals of the Royal Academy 1893, p.176).

Of the factors that shaped Viñes' austere character and scientific discipline, one cannot underestimate the role played by his religious order, the Society of Jesus, which he joined as a very young man. Viñes is best understood within the historical context of the Jesuits, renowned for their strict discipline and devotion to learning.

The Jesuits have a tradition as missionaries, educators, and men of science going back to the founding of their order in 1542. To this day, at least one-quarter of Jesuits are directly involved in education, many in higher education. The first Jesuits arrived in Havana in August 1566 as part of a commission sent to the New World by Philip II of Spain. After being expelled from Spain and its colonies in 1767, the Jesuit order was suppressed by Pope Clement XIV in 1773.

Pope Pius VII restored the order in 1814, but the Jesuits did not return to Cuba until the second half of the nineteenth century, when they decided to establish a college in Havana. The location they chose was the site of what had been the Convent of Belén and was then serving as a Spanish Army barrack.

The first class of the Royal College of Belén began on March 2, 1854, with 40 students and a school day that stretched from 6 AM to 6 PM. Shortly after, there was a dire need to establish a laboratory where students could apply the physical and natural science theory they were learning in the classroom. It was this practical need, and not the threat of hurricanes, that first drove the College to establish its Observatory.

The Belén Observatory began taking meteorological observations on March 1, 1858 at 8:00 AM. From that point on, observations were made four times daily: at 8:00 AM, 12:00 PM, 4:00 PM, and 8:00 PM. The observations included barometric pressure, temperature, and rainfall. A month later readings of vapor pressure were added. In September of that year maximum and minimum temperatures were added to the suite of observation, and by October cloud observations and wind direction were added to the noon observations.

Benito Viñes took over as director of the Belén Observatory twelve years later. He gave it his heart and soul during his tenure there, drawing on the legacy of stoicism and perseverance of his predecessors. He made the best of the meager resources at his disposal, with a selfless focus on contributing to the welfare of humanity.

A Look at His Work
and the Evolution of
His Scientific Thought

3

Once a cyclonic movement begins to develop in the atmosphere, an observer should conduct a thorough examination of all the phases of its evolution and examine his conjectures about what is observed, with the goal of not only advancing science one more step, but also of being able to resolve the various problems that emerge, making use of whatever observations are within reach, his own or those of others, and basing them on known laws. (Viñes 1877, p.36)

Let us now examine the results of Benito Viñes' labor. This being a biographical work, it is not the best medium through which to do a thorough assessment of the scientific value of each work he produced, but we will point out salient aspects of his contributions, and try to categorize them so as to gain a better understanding of the areas of knowledge in which he conducted his professional work.

There are three places where Father Viñes' entire body of work is catalogued. These are "Catalogue of the work and writings of Fr. Viñes," on the back cover of the book *"Investigaciones*

Relativas" written in 1895 by Father Lorenzo Gangoiti, S.J., which lists 15 publications, among which are the annual bulletins of the Observatory, and the large-format albums of press clippings mentioned in chapter 1.

The second compilation of Viñes' work, this time listing 14 items, was published nine years later by Fr. Gutiérrez-Lanza (Gutiérrez-Lanza 1904). The third appeared in the Annals of the Royal Academy and comprises 12 documents, including presentations made at that forum by Viñes and the serial publication *Apuntes Relativos a los Huracanes de Las Antillas…(Notes Concerning Hurricanes in the Antillles…)*, published in 1895. Since none of these gives an accounting of Viñes' complete opus, we have compiled a new list of his scientific work that excludes his press releases, far too numerous to include in this book. Viñes' scientific output can be put into three general categories:

1) Presentations made by Viñes on a single theme, whether published or not.

2) Books about a variety of topics.

3) Scientific articles, forecasts and various types of information particularly aimed at mass distribution by the press.

The contents of the first two categories are described below in chronological order:

1. **Magnetic perturbations and the aurora borealis of 4 February 1872**

   1872—geomagnetism—monograph—13 pp.

This is from a presentation made to the Royal Academy of Medical Physical and Natural Sciences of Havana, published in two parts in Volume 9 of the Royal Academy. This was likely the first contribution to the Academy by Viñes and presented data recorded at the Observatory during an aurora borealis observed in Havana as well as North America, Europe, and parts of Asia and Africa. Viñes opines that this phenomenon was electric and magnetic in nature, and that it occurred in the upper atmosphere and not in outer space as was generally believed at the time.

It is now a scientifically accepted fact that the polar aurora is not a phenomenon that occurs outside the outer limits of our atmosphere, but is rather a phenomenon that occurs within the atmosphere... It is in addition an eminently electrical phenomenon, and as such it is intimately connected with Earth's magnetic field, which it profoundly affects. (Annals of the Royal Academy 1872, p. 117)

Reading through this monograph, we learn of the close relationship between Viñes and Father Perry, who was in England at the time. In another paragraph, Viñes writes:

Fr. Perry, director of the Stonyhurst Observatory, wrote me that at the height of the event he was also unable to obtain readings from the bifilar because the reading had gone off the scale. (Annals of the Royal Academy 1872, p.119)

It appears Viñes believed there was a close relationship between magnetic disturbances and local meteorological changes, an opinion he shared with Father Secchi and that could be summarized in three principles: "No storm occurs without a previous or simultaneous disturbance," "magnetic perturbations occur one, two, or even three days prior to a local storm," and "large perturbations can serve as an early warning sign of stormy weather."

Note the priest's interest in finding a scientific foundation for making weather forecasts and willingness to work toward that goal. It is possible that in articulating the third principle, Viñes was seeking to determine if there might be a link between magnetic disturbances and hurricanes.

2. **Magnetic perturbations in relation to other meteorological elements during the months of July to October**

1872—geomagnetism—monograph—10 pp.

This was another work presented at the Academy of Sciences and published in Volume 9 of its Annals, in five separate installments. In it Viñes speculates about the relationship of magnetic disturbances and thunderstorms in particular. He states: "Are these mere coincidences, or is there a link between magnetic and meteorological phenomena? The accumulation of more observational data over time should help us answer those questions." (Annals of the Royal Meteorological Society1873, p.241). This monograph is accompanied by data tables of meteorological and magnetic observations.

3. **The storm of 6 October 1873**

This is a scientific analysis about the "stormy weather" experienced in Havana during that day, which Viñes originally did not attribute to a hurricane. In the Gangioti compilation, this work's title originally appeared as *"The storm of 6 October 1873 was not a hurricane."* Viñes presented this paper at the Royal Academy session of 12 October 1873 and it was published in Volume 10 of the Annals of the Royal Academy in four parts.

Viñes begins this presentation by describing the weather observations taken during the period of September 28 to October 6 of that year and attributes them to the arrival of a cold front (a *norte* in local parlance). Viñes observes that in "these regions" (referring to Cuba):

> *After the strong current of S. winds ends, having so to speak exhausted itself, the wind shifts to NW in a short period of time. Hence the popular rhyme, which I have always found to be true: "Sur duro, Norte seguro" (Strong South, North no doubt) (Annals of the Royal Academy 1873, p. 175).*

Viñes bases his analysis in the characterization of polar and tropical air masses and attributes the origin of these types of storms to the clash of these air masses. He further elaborates this scenario to attempt to elucidate the origin of hurricanes:

The perpetual struggle between these two currents, polar and tropical is, in simple terms, the basic cause of all atmospheric changes, including hurricanes, which are no more than a single episode in the ongoing struggle between these two great powers. (Annals of the Royal Academy 1873, p. 175)

Viñes shares his conjecture of what a hurricane would look like if he were to position himself outside the atmosphere and obtain a synoptic view of this phenomenon:

If we were given the ability to observe this phenomenon from the top of our atmosphere, the polar and tropical currents would appear as two very broad rivers, whose courses alternate between placid pools and rapids, both rivers flowing in different directions and occasionally overflowing their banks; while the hurricane would appear as a black dot in the midst of this great atmospheric ocean, a disk of small size, and sinister aspect, made of tightly packed black clouds spinning at a dizzying rate... (Annals of the Royal Academy1873, 176-177)

Ninety years later, satellite pictures would show meteorologists for the first time a strikingly similar scenario.

In this work, one can already discern the origin of ideas that would later be more fully explained in his *Apuntes Relativos*. In this work about the October 1873 storm, he arrives at tentative conclusions, which he enumerates in a list of 22 items. Among these are "precursor signs" of a hurricane, which he eventually incorporated in his hurricane forecasting methodology. These include:

15a. Full cloud cover, low fleeting clouds against a dark leaden background, at night occasionally accompanied by a kind of sinister phosphoresce, and heavy downpours that increase in intensity as the vortex approaches.

16a. The terrifying aspect of the sky typical of a hurricane is the same regardless of wind direction...(Annals of the Royal Academy 1873, 176-177)

These and other ideas that appear in this document, the third of his scientific contributions, show the development of his thinking and his interpretation of the observations he was making, which culminated in his *Investigaciones Relativas a la Circulación y Traslación Ciclónicas* (Investigations Concerning Cyclonic Circulation and Movement).

4. Curious phenomenon observed in Havana during the magnetic perturbations

1873—geomagnetism—monograph

According to Gutiérrez-Lanza, this work appeared as an appendix in the annual summary of observations for 1873 published by the Observatory. Unfortunately, I have been unable to find this document.

5. Regular and irregular variations of the barometer in Havana from 1858 through 1871

1874—meteorology—monograph—21 pp.

This is an analysis of fourteen years of atmospheric pressure data at Havana recorded since the founding of the Observatory. It appeared in the Annals of the Royal Academy in four parts in Volume 11, published between 1874 and 1875. Viñes had to perform long and tedious calculations in the calculation of these data.

He states:

> *Although I was aware of the arduous and onerous task ahead*
> *of me when I began this work, I must confess that once I*
> *started, it would have been negligent of me to abandon the*
> *project. I was sustained in this effort by its potential usefulness,*
> *a goal that would be of immense value to me.* (Annals of the
> Royal Academy 1874, p. 259)

Viñes was evidently trying to obtain a statistical basis from
which to deduce regularities in the behavior of atmospheric pressure and perhaps use in forecasts: "Only this hope can sustain the
drive necessary to carry out these long and monotonous tasks"
(Annals of the Royal Academy 1874, p. 255).

Viñes began by calculating the mean values of readings taken
every two hours between 4:00 and 22:00 h for each month, which
he then used to find the characteristics of the atmospheric tide.

> *In each and every curve derived, one can find two maxima*
> *and two minima. The first minimum occurs between 2 and*
> *4 AM, the first maximum between 9 and 10 AM. The second*
> *minimum between 3 PM and 4 PM closer to the latter, and the*
> *last maximum near 10 PM.* (Annals of the Royal Academy
> 1874, p. 259)

Later in this monograph he refers to the advantages he expects
to see with the installation of Father Secchi's meteograph: "Once
we can make use of Father Secchi's meteograph, which is currently
being installed in this Observatory, I will be able to make these
types of observations more easily."

Viñes also noted changes in the shape of the barometric pressure graphs from month to month:

The peaks are sharper in December, January, and February, and the differences between maxima and minima in these months are considerable. Starting in March, these differences diminish rapidly and the curves dampen until June, July, and August, and begin to increase in amplitude again in the months of September, October, and November. (Annals of the Royal Academy 1874, p. 307)

Further down, Viñes references Kaemtz, Humboldt, and Marie Davy and summarizes his conclusion of a general character. We mention the second of these, which we consider the main one:

Of all the barometric observations:

a) *The diurnal one is very regular in Havana, as seen in a decade's worth of data, or even a single day's.*

b) *The diurnal curve consists of two maxima, at 10 AM and PM and two minima around 4 AM and PM.*

c) *The morning maximum and the evening minimum are the most pronounced.*

d) *There is a seasonal variation in the times of maxima and minima.*

e) *And finally, the diurnal variability always produces a measurable effect, even during major atmospheric commotions.* (Annals of the Royal Academy 1873, p. 370)

6. **Notes concerning the hurricanes in the Antilles in September and October 1875 and 76**

> 1877—meteorology/hurricanes—text of a
> general nature about this subject—256 pp

This is the most extensive work by Benito Viñes. It includes, first of all, a complete set of presentations made at the Royal Academy and published in Volumes 14 and 15 of its Annals. The collection was published as separate book, which gave Viñes international exposure. In the first part of this book, Viñes recounts aspects of the site visits he undertook to document the impacts of these hurricanes.

I will not go into the anecdotal aspects of these journeys, as they have been discussed in chapter 1. In the part of the book dealing with his conclusions, however, Viñes emerges as an expert in the field. He organizes his work thus:

> *In order to proceed with the best order and clarity possible in analyzing the results…based on the data gathered so far… I shall divide them in the following four categories:*
>
> *I. Determination of the path of the hurricane.*
>
> *II. Investigations related to the body of the storm.*
>
> *III. Examination of the phenomena observed during its passage and its aftermath.*
>
> *IV. Theoretical issues and practical applications that can be deduced from the observations.* (Viñes 1877, p. 52)

Although chapter 2 of this book is of great interest as far as the description of the hurricanes experienced in Cuba in 1875 and 1876, it is chapter 3, which is focused on the study of the hurricane as a phenomenon and the effects it produces "as it approaches and recedes," that constitutes the most important part of this book.

Chapters 3 and 4 provide the foundation of the forecasting principles that came to be known as the "Viñes laws" that were used for hurricane forecasting for many years afterwards.

Of interest is his study of optical phenomena and the subtle changes in color and appearance of the atmosphere he had noted prior to the passage of a hurricane, especially the cloud evolution associated with them, particularly cirrus clouds, which he called "plumiformes" (feather-like in shape).

Viñes used these cloud observations as a way to estimate the geographical location of the vortex from a great distance, a matter of great practical utility to ships' captains and others at risk of being in the path of the storm. Chapter 4 is dedicated to providing practical advice to mariners, and contains recommendations on the best strategies to follow to either ride out a hurricane or avoid it altogether, all based on application of the "laws."

This book discusses several other aspects of well-defined hurricanes, among them the patterns of rainfall observed during times when the path of the hurricane curves, the so-called "pumping" effect he had seen on barographs during the passage of a hurricane, and the combined effects of water and wind in causing these natural disasters. He also discusses in detail the differences between the flooding produced by the storm swell as it reaches the coast and the storm surge of the hurricane proper.

Viñes devotes several pages of this book to the hydrological aspects of heavy rainfall associated with tropical storms. He addresses a topic that had not received much attention before: the role played by the limestone landforms of the plains of western Cuba, where its numerous sinkholes and caverns (known locally as "casimbas" or "sumideros") are identified by the author as natural "wells" that play a major role in drainage by rapidly absorbing much of the rainfall and reducing runoff. He warns against interfering with this natural drainage by filling or otherwise altering these features of the landscape:

> It was then that the waters began to rise and tumble uncontrolled over the sugar mills. This was due in no small part to the fact that the opening of many of these sinkholes had been obstructed and they proved insufficient in absorbing the sudden surge of water. (Viñes 1877, p. 219)

One must add that this problem continues to this day. A recent example of the consequences of human interference with natural drainage was the flooding of the runway of the José Martí International Airport in Havana during Hurricane Frederick's passage south of Cuba in 1979.

In this book, Viñes was one of the first to advocate for the construction of a canal to prevent future floods, anticipating what eventually became the "Roque canal" that was built years later:

> ...canals of sufficient carrying capacity should be built connecting lagoons to one another and eventually to the sea to prevent ponding of rainfall. Enacting this measure should make future floods less fearsome, and since these are brief events, they could even become beneficial to agriculture.
>
> The cost of this enterprise is not such to compare to the much greater losses incurred in even a single flood, since in the Santa Rita sugar mill alone the losses from the 1876 event exceeded 10 thousand boxes of sugar, whose value would exceed the cost of construction of the canal and would bring incalculable benefits. (Viñes 1877, p. 224)

This book is undoubtedly the most comprehensive Viñes produced. It brought together his research work and put it to use in concrete ways for the benefit of society. It continues to be an indispensable reference to anyone wishing to study the history of natural disasters in Cuba, in addition, of course, to its being a contribution to the study of tropical cyclones.

7. **The hurricanes of 7 and 19 October 1870**

   1878—meteorology/hurricanes—monograph—20 pp.

This pamphlet was published as an appendix to *"Observaciones Magnéticas y Meteorológicas del Año 1870"* (Meteorological and Magnetic Observations of the Year 1870). The eight-year delay in the publication is accounted for by the time-consuming numerical calculations required to prepare the data tables.

This work describes an in-depth study of the features and effects of these hurricanes. In the paragraph quoted below, there is evidence that Viñes was able to gather retrospective data about these events during the trip he made to Matanzas in 1876 by interviewing witnesses of these events from six years before:

> Some of the data on which we base our analysis comes from reliable witnesses we interviewed, newspaper accounts from the capital, observations provided by the Navy, observations from our own Observatory and from our in situ inspection of the areas most affected by these events... (Viñes 1870, p. 3)

Elsewhere in this publication, Viñes cites some of the theories advanced by William Redfield in 1831, and brings to bear the thoughts of other authors about the internal mechanism of hurricanes. Viñes postulates that these hurricanes must have reached 250 miles in width "from Sancti Spiritus to Pinar del Rio", and surmises that the maximum winds near its center reached a speed of 40-45 meters per second "if one considers the devastation caused in the vicinity of Matanzas" (Viñes 1870, p. 15).

This text also examines the hydrological events that were observed in the area during and after the passage of the hurricane. In particular, Viñes takes note of the distinct phenomena of storm surge (as well as the nearby accompanying decrease in coastal sea levels), and the coastal flooding caused by storm swell, all lumped together in popular parlance as the "ras de mar."

> Another, more terrible event took place in Matanzas: the flooding caused by the sudden rise in sea level impeding the flow of the rivers to the ocean... This sea level rise in Matanzas was accompanied by a sea level depression in Cárdenas (also located on the north coast of Cuba, 100 miles east of Havana) as well as in Batabanó (on the south coast of Cuba, directly south of Havana and approximately 50 miles SW of Matanzas), where

the water retreated from the beach. This coastal event, which generally accompanies hurricanes, can be explained by the reduction in pressure near the vortex, acting in concert with the high surf generated by the winds. (Viñes 1870, 14-15)

This publication marks the place where Viñes makes his first, though vague, reference to observation of cirrus clouds. In it he mentions that during the period 1-3 November of that year, another hurricane appeared to the west, affecting Havana with "low barometer and a few gusts from the SE, having been preceded by cirrus and accompanied by nimbus and dense cumulus from 31 October to 4 November."

It is possible that these lines were edited well after they were written, but there is no doubt that Viñes was already a careful observer of clouds as early as 1870.

8. **Absolute determinations of the declination, inclination and horizontal terrestrial magnetic force from observations made at the Observatory of the Royal College of Belén in Havana.**

     1887—geomagnetism—monograph—5 pp.

This publication is a summary of geomagnetic observations made at the Observatory in the years 1885 and 1886 and appeared in the Annals of the Academy in Volume 24 (Annals of the Royal Academy 1887, p. 81).

9. **Visit to the Vuelta Abajo region by Reverend Father Benito Viñes, Director of the Observatory of the Royal College of Belén, and Mr. Pedro Salteraín, Mining Engineering Expert of the Island of Cuba, on the occasion of the strong earth temblors that occurred during the night of October 22–23, 1880.**

     1880—geology/seismology—monograph

This work appeared as a supplement to an edition of the *El Triunfo* newspaper and was a report on a site visit made by Viñes and Salteraín to several towns damaged by two strong earthquakes that occurred near midnight on the night of January 22 and 23, 1880 on a fault running along Cuba's best tobacco growing region. According to a recently published catalogue,[38] the earthquake epicenter was located near the town of San Cristóbal, in the Province of Pinar del Río west of Havana. The strongest quake is estimated to have reached an intensity of 8 on the MSK[39] Scale and was followed by 65 aftershocks. Surprisingly, given its magnitude, the event caused only three fatalities, but it was felt widely, even as far as the Florida Keys.

In his book *"Fr. Viñes and his Scientific and Humanitarian Works"* Gutiérrez-Lanza points out that this trip was primarily intended to reassure the local population, and that they visited the area around the epicenter of the quake, where some deaths were reported. In his *"Apuntes Históricos acerca del Observatorio de Belén"* Gutiérrez-Lanza points out that;

> *[The trip's] purpose was to study the land of the Western Province from a geologic and volcanic perspective. A report of this trip, which is not part of the catalogue of his [Viñes'] work....*
> (Gutiérrez-Lanza 1904, p. 22)

---

[38] Álvarez, L., and Coauthors, (1999): An earthquake catalogue of Cuba and neighbouring areas. The Abdus Salam International Centre for Theoretical Physics, Miramare–Trieste. Internal Report IC/IR/99/1.

[39] The Medvedev–Sponheuer–Karnik (MSK) Scale measures the severity of ground shaking in an earthquake. The MSK Scale, used in Russia and other countries, is similar to the Modified Mercalli Scale used in the United States for the same purpose. An MSK 8 earthquake is described as: "Many people find it difficult to stand, even outdoors. Furniture may be overturned. Waves may be seen on very soft ground. Older structures partially collapse or sustain considerable damage."

10. Magnetic perturbations and their relation to Nortes40 and principal atmospheric changes

1881—meteorology/magnetism—monograph

This work is the compilation a series of monthly reports in which Viñes compares changes in the local weather to significant changes noted in the magnetic observations recorded by instruments in the Observatory. Though published in 1881, the data analyzed were obtained in 1874 and 1875.

In Gutiérrez-Lanza's compilation there appears, in addition to this title, a similar one entitled *"the Nortes of the Gulf and their relation to magnetic perturbations."* Gutiérrez-Lanza apparently considers these to be two different works, though they both refer to the same series of monthly reports. Viñes does not make mention of concrete results or conclusions, but rather limits himself to a side-by-side listing of magnetic and meteorological fluctuations, noting along the way coincidences in large fluctuations of the meteograph and the bifilar magnetometer.

Viñes also compares these observations with the presence of other upper atmospheric phenomena like the aurora borealis. It was noteworthy to us that Viñes made no mention on this work to the cycles of solar activity, a subject that was amply discussed by Father Angelo Secchi in his treatise *Le Soleil*, which we assume was well known to Viñes.

11. The Storms of Cuba

1886—meteorology—monograph

This work is focused on attempting to uncover links between Cuban storms and weather events that were reported in North America in the year 1875. It was published initially in the annual report of the Observatory of Belén and later summarized in Volume 23, pages 334–341 of the Annals of the Royal Academy.

12. Observations of the Transit of Venus, made at the College of Belén in Havana, 6 December 1882

1882—astronomy—monograph—8 pp.

In this brief monograph, Viñes describes the observations made as Venus crossed the disk of the Sun during that date. This work also describes the Cooke telescope, then newly acquired by the Observatory for use in astronomical observations. This work was limited to describing key event in the transit of Venus, such as the times the transit began and ended.

13. Unusual trajectory of the disastrous hurricane of 4–5 September 1888

1888—meteorology/hurricanes—monograph

This work appeared as an appendix to the annual publication summarizing the meteorological observations made that year. It describes details of a hurricane that took a track unlike those of its recent predecessors. It made landfall on the north coast of Las Villa province, just east of Sagua La Grande41 and left Cuban territory toward the Yucatan Channel near Mantua, Pinal del Rio Province, in far western Cuba. The unforeseen path of this hurricane surprised Viñes and the residents of the westernmost province who were totally unprepared for this event. This monograph is primarily focused on the trajectory of the hurricane, which did not conform to Viñes' predictions.

14. Cyclonoscope of the Antilles

1888—meteorology/hurricanes—instrument

This instrument designed by Viñes was based on his "laws" and was intended to be of use to mariners at sea in the Caribbean or Gulf of Mexico area. It was essentially   a kind of circular "slide rule" consisting of a fixed part, inside of which was centered a concentric disk of smaller diameter. By rotating the movable inner disk, one could determine the direction of wind and cloud motion at the observation location. The outer circle was engraved with the 16 compass directions. In the inner, rotatable disk, one would line up, in order, wind direction, and the direction of low clouds, altocumulus, and cirriform clouds, including the "plumiform" cirrus whose movements had captured his attention from the start. With this instrument, one could then estimate the direction of the vortex by establishing the relative motion of the various meteorological elements.

This instrument, in spite of its simplicity, made practical use of eighteen years' worth of careful cloud observations. It is still surprising that it did not come into wider use by mariners over the years.

15. Cyclononephoscope of the Antilles

> 1888—meteorology—hurricanes—instrument
> and instruction booklet

This is basically the same instrument described above, this time relabeled to be more etymologically correct with the addition of "nephos," Greek for "clouds." The accompanying instruction booklet not only described the proper use of the instrument, but also discussed the theoretical basis on which it was constructed.

16. Investigations of the circulation and movement of hurricanes in the Antilles

> 1893—meteorology/hurricane—technical text

This is Viñes' "scientific testament," as described by several authors. The manuscript was written at the invitation of the organizing committee of the Meteorological Congress held that year in Chicago-clear evidence of the international reputation that

Viñes had acquired as an expert in tropical meteorology. Published posthumously, this work describes the rules he had derived for monitoring the track of tropical cyclones, based on his statistical analysis of 38 such phenomena.

The procedure for tracking the position of cyclones described in this book starts with recording the surface wind direction at the location of observation, and then proceeding upward, noting the winds at various levels, as indicated by cloud motion vectors, all the way to the highest cirrus clouds that could be seen. Then, using as a reference one or more of these wind vectors, the observer could deduce the compass direction of the storm vortex with respect to his location, using the "cyclononephoscope" based on Viñes' laws. Viñes, in keeping with the custom of his time, used the term "laws" to refer, improperly, to what in effect is the statistically most commonly observed behavior of tropical storms.

Viñes structured this book in two parts. The first focused on cyclonic circulation (around the vortex), where he described the horizontal and vertical structure of tropical cyclones. The second part was concerned with the average cyclone trajectories at various times during the hurricane season including the latitude at which their path typically changed direction.

In the second chapter of this book, under the subheading "Principal Anomalies Observed," Viñes pointed out individual cyclones that deviated from these "laws" either in structure or trajectory, or both, or whose behavior he could not explain.

Later in this book, Viñes enumerates a set of rules for ship captains to follow if they find themselves in the vicinity of a tropical cyclone, a particular concern of Viñes.

This work should be considered in its historical context. It is worth noting how the ideas presented here by Viñes diverged from those authors whose theories Viñes initially thought applicable to Cuba, such as William Piddington in his "Law of Storms," or those proposed by William Redfield. Using the experience gained during his years at Belén Observatory, Viñes departed from the notions he brought with him to Cuba. Take this example, where Viñes discusses his observations of low-level convergence near the vortex of the storm and reaches a different conclusion from the above-cited authors:

> It is, of course, clear that the law I just stated is of major importance in the theory of cyclones, but leaving theoretical considerations aside for the moment, I can add that it is no less productive as far as practical results and general utility than the so-called LAW OF STORMS as far as cyclonic rotation is concerned. In brief, the LAW OF STORMS only takes into account the surface-level circulation of the wind, which, in addition to not being circular as assumed by that law, also exhibits the most variability in the angle made by the (local) wind direction with respect to the vortex, when compared to the other (higher) currents, and it is the one most subject to irregularities and local influences.
>
> On the contrary, THE LAW OF CYCLONIC CURRENTS AT DIFFERENT ALTITUDES takes into account the wind currents at all levels of the atmosphere that can be observed, accounting for the fact that the lower currents experience convergence, and modifying the Law of Storms to better fit the observations of air currents at different altitudes. This law, therefore, **describes and makes evident the cyclonic movement of air in the atmosphere....** (Viñes 1895, 19-20)

There is evidence in this book of a basic change in Viñes' ideas about the structure and dynamics of hurricanes, a transition he had gradually begun manifesting in previous works and was now fully articulating. In his initial work, Viñes had conceived of hurricanes as being essentially rotating disks. By the time he wrote *"Investigaciones Relativas..."* his view had evolved:

The cyclone should not be viewed as a mass of air that gyrates as a whole body, as would a wheel in motion responding to a push. The cyclone is better viewed as a partial vacuum, that propagates and is constantly renewed along its path, pulling in air from all sides surrounding its base, setting it in gyrating motion, and expelling it in diverging currents in its upper parts. (Viñes 1895, p. 76)

This book summarizes Viñes' life work. It serves as a complement to his other great achievement: his compendium of hurricanes in *Apuntes Relativos a los Huracanes de las Antillas* (Notes Concerning the Hurricanes of the Antilles). Special mention should also be made of his contributions to the Observatory publications containing the meteorological observations made there, summarized by season and year. He started this ongoing task in the decade of the 1860s. The oldest of these bulletins that we have found is one entitled "Magnetic and meteorological observations made by the students of Belén College of the Company of Jesus in Havana," covering the period 1858–1859, before Viñes' arrival.

As far as the data bulletins published under Viñes' supervision, one can say that they are characterized by their high quality and careful editing, as well as their progressively longer delays in getting published, no doubt due to the tremendous workload of the small workforce that produced them. One can only imagine how arduous a task it was to calculate and error-check the great mass of observational data by hand. According to Gutiérrez-Lanza, by 1893, the year Viñes died, he had only managed to publish the data bulletins covering the periods 1870–1875 and 1885–1889. Not much more could have been asked of him. As an example of how tedious and prolonged and enterprise this was, it took Father Lorenzo Gangoiti, Viñes' successor, until 1901 to prepare for publication the data obtained in 1877.

Viñes attached great importance to producing the meteograms and other graphs that he began using to supplement the data tables in these publications. To ensure the quality of these graphs, he worked directly with the lithographic "stones" that were used in the printing process:

> *It is astounding to learn how that man could handle the daily work of the Magnetic-Meteorological Observatory, where it was necessary to make and record observations every two hours… prepare the monthly publications and personally produce the lithographic plates, and attend to any number of consultations and host visitors to the Observatory….* (Gutiérrez-Lanza 1904)

Other interesting items are the press articles edited by Benito Viñes. Some of these are directed at educating readers on scientific aspects of meteorology. It appears that Viñes found it necessary to impart to the general public a basic understanding of hurricanes so that they would make more informed uses of his forecasts.

An example can be seen in an article that appeared in *La Voz de Cuba* on October 17, 1879 that also appeared in *Diario de la Marina and El Triunfo*. This article explains in significant detail how the atmospheric pressure varies in hurricanes of various diameters. His explanation of these concepts is clear, and free of jargon and unnecessary details.

The year 1882 began for Viñes with what must have been a carefully planned journey to Europe. The purpose of the trip this time was to improve the technical capacity of the Observatory by acquiring better quality and more up-to-date instruments. As part of the trip he planned to meet researchers that could advise him of the criteria to use in their selection and acquisition.

I have uncovered no explicit references to the planning of this trip, nor any details of his itinerary. However, in a note dated April 4, 1882 bearing the seal of the College, one can read in Viñes' distinctive handwriting:

It being necessary for me to leave for Spain with the next mail ship and from there book passage to England, where I plan to stay a few months to tend to matters related to the Observatory. I would like to inform you, and through you the Academy, that I am glad to offer my services while there in any matter that might be of utility.

May God keep you,
Havana, 4 April 1882 (Historical Archive of CEHCYT)

The skimpy information we had at our disposal about this trip motivated us to do a careful search of the newspapers published in 1882. With good luck and after much effort, we found the following note in the *Diario de la Marina:*

PORT OF HAVANA
DEPARTURES

Postal steam ship "Mendez-Nuñez" en route to Puerto Rico on 5 April

This was followed by the manifest. Among the passengers listed, as we expected, was Viñes' name. This means that, after leaving Havana on 5 April, he would have arrived in Europe, by way of Barcelona, around April 20 to April 25. After twelve years' absence, he was able to see his native land briefly, and then went on to England, France, and Belgium to meet with the leading meteorologists and astronomers of the day.

In England, he made an extended visit to fellow Jesuit Fr. Perry (1833-1889), director of the Stonyhurst College Observatory[42] in

---

[42] The Stonyhurst Observatory is part of Stonyhurst College, a Jesuit institution. Daily temperature records have been taken there since 1846 and constitute one of the oldest continuing temperature records in the world.

Lancashire. Father Perry trained Viñes in the use of the new instruments the latter had acquired on the trip so they could be properly installed once they arrived in Cuba. During his visit to England, Viñes also attended a course given by G.M. Whipple, a renowned astronomer who was at the time director of the Kew Observatory, located within the Old Deer Park, in Richmond, near London.

Viñes returned to Cuba in mid-1882 with several of the best and most modern meteorological and magnetic instruments then available. The August 29 edition of the Havana daily *El Avisador Comercial* published a note that read:

WELCOME

We extend a cordial welcome to the Rev. Fr. Don Benito Viñes on the occasion of his return to this city, where he arrived in the postal steamer from the Peninsula. From what we are told, the illustrious director of the observatory of the Royal College of Belén has acquired in Europe various meteorological instruments of the best quality, as would be required of this observatory, considered in that part of the world to be on a par with the best in the world.

A complete list of all the scientific instruments Viñes brought with him from Europe has not survived, but the suite included the following astronomical instruments: three theodolites, a sextant, a chronograph, a chronometer and, most important, a high quality refracting 152 mm (f=13.7) equatorial mount telescope built in England by Cooke & Sons. The magnetic instruments included three declinometers, three magnetometers and three compasses, with distinct specifications pertaining to each instrument. The meteorological instruments comprised five barometers, various thermometers and psychrometers of various types, an evaporimeter, two anemometers and wind vanes, several nephoscopes and an actinometer. Most of these instruments were calibrated against reference instruments at Kew Observatory.

During 1882 and 1883, Viñes was a participant in the activities of the International Polar Commission, which involved taking supplementary observations at the Belén Observatory during the period 1 September 1882 to 31 August 1883. These included nine additional observations during odd-numbered hours and readings of the magnetometer and declinometer every five minutes day and night for days 1 and 15 of every month, synchronized with the Gottinger, Germany, mean time. The inclusion of Viñes in these scientific activities, of course, was a testament to Viñes' growing international reputation.

An Exceptional Observation of the Transit of Venus in 1882

Among the astronomical events of greatest interest to nineteenth-century science were the planetary transits, when one of the two planets with an orbital radius smaller than Earth's, Mercury and Venus, passes in front of the disk of the Sun as seen from our planet. The precise alignment needed for these events to occur makes planetary transits very infrequent, but they present opportunities for astronomers to make accurate measurements of the distance from the Earth to the Sun and to better determine planetary orbits.

Consequently, the rare transits of Venus were significant events in nineteenth-century astronomy. The first historical observation of the transit of Venus was made in England on November 24, 1639 (corresponding to December 4 in the Gregorian calendar now in use) by the astronomer Jeremiah Horrocks (1619–1641). Since then, many observatories have organized expeditions to observe these transits at different latitudes.[43]

---

[43] The transit of Venus has occurred twice since 1882, in 2004 and 2012. The next one will occur in 2117.

Viñes and his team at the Royal College of Belén seized the unusual opportunity to observe the transit of Venus from Havana on December 6, 1882.[44] By this time, Viñes had at his disposal the state-of-the art instruments he had acquired in his recent trip to Europe, among which were an equatorial-mount 152 mm refractive telescope manufactured by Cooke & Son, established in Hull, England, in 1863. At the time, it was the only telescope of the type in Cuba and one of the best in Latin America.

To carry out this task, the use of a good telescope was not the only requirement; accurate time measurements were also needed. To that end Belén Observatory had also purchased an accurate chronometer that had been carefully calibrated with the Jesuit-run Stonyhurst Observatoy in the United Kingdom, as well as a very accurate pendulum chronograph manufactured by the firm Howard & Co. of Boston, which had cost the extraordinary expense (for the time) of 600 dollars.

The Belén Observatory also owned a Troughton theodolite, of the same model that Joseph Perry, S.J., used in Madagascar to observe the same 1882 transit of Venus. To determine the exact location of the Belén Observatory, Viñes used a Jones sextant equipped with an "artificial horizon." In summary, the Observatory of Belén was well prepared for the 1882 transit, using all the internationally accepted standards of astronomical practice and the best equipment available in the last third of the nineteenth century.

To prepare for this event, Benito Viñes assembled a work group that included two other Jesuits working at the College: Fathers Pedro Osoro, S. J., and Bonifacio Fernandez Valladares, who had had prior experience in the areas of meteorology, geomagnetism, and astronomy. Also helping in this effort was the layman Clemente López, who was employed by the College.

---

[44] Viñes, B. (1887): "Observación del Paso de Venus, Hecha en el Real Colegio de Belén el 6 de Diciembre de 1882", *Anales de la Real Academia de Ciencias Médicas, Físicas y Naturales de La Habana*, t. 24 (pp. 74-81), Imprenta "La Antilla", La Habana.

Each member of the team had an assigned task. Viñes made visual observations through the telescope's eyepiece, while Fr. Osoro worked the chronometer and called out the time to ensure exact observations. The exact time, together with a description of the observations made by Viñes, was recorded by López. Finally, Fr. Valladares acted as data quality manager to prevent inadvertent errors creeping into the data.

In preparation for the transit, Viñes had devised a set of shorthand words to be used at key times during the event, in order to minimize latency in recording the data and minimize the risk of confusion. Viñes delved in depth into the theory behind the observations of the transit, starting with a general announcement published in Washington that had been sent to various observatories and to American expeditions sent to observe the event. Similarly, two Spanish commissions had sent instructions for use in its American colonies, and the British had sent an instruction manual for their expedition in Madagascar, all for the purpose of observing the transit of Venus at different latitudes.

Viñes also had at his disposal the 1882 Nautical Almanac of the San Fernando Observatory, as well as the article about the upcoming event that had appeared in the 22 June 1882 edition of the journal Nature. All of the planned transit of Venus observations had been coordinated in advance during the International Astronomical Congress in 1881.

The team readied the instruments during the early morning hours of December 6, 1882, installing protective filters over the telescope's eyepiece, an indispensable item, since a careless mistake can produce serious injuries to the eye of the observer or even cause permanent blindness. Fr. Viñes chose a lens that magnified the image by a factor of 150.

Since the telescope had only recently been purchased, the Observatory did not yet have a dome to house the telescope, so the observing team set up its tripod on the roof of the building directly above one of its central support columns, so as to minimize vibrations caused by people walking about within the building. Observers were protected from glare during their observations by strategically placed canvas sheets.

December 6 dawned with some cloud cover, but fortunately the clouds dissipated in short order. One limitation of the available equipment-the lack of an accurate calibrated circle that would have allowed fixing the angular position of Venus with respect to the Sun's disk-prevented Viñes' team from recording the exact location when Venus began its transit over the circumference of the solar disk. The transit was recorded to have begun at 8h 43m 34s and ended at 14h 39m 40s local time.

The Havana observations of the transit of Venus were successful. The Royal College of Belén's team was able to obtain a good record of three contacts[45] of the transit, including a phenomenon known as the "black drop"[46] and to observe a halo around Venus caused by its atmosphere. The data obtained were sent to Washington, Greenwich, Paris, and other astronomical observatories in various parts of the world. The efforts of Viñes and his team, therefore, made them pioneers of astronomical observations in Cuba.

---

[45] As the smaller disk of Venus passes over the disk of the Sun, there are four "contacts": the first when edges of the two disks touch at a point as Venus begins its transit from a position outside the disk of the Sun, the second when the two disks touch at a point as the disk of Venus is entirely within the Sun's disk, and similarly in reverse order with the third and fourth contacts, as Venus exits the Sun's disk.

[46] Just after second contact, and again just before third contact during the transit, a small black "teardrop" appears to connect Venus' disk to the limb of the Sun, making it impossible to accurately time the exact moment of second or third contact.

A Large Fireball Over Havana

Fireballs (large meteors also known as bolides) have always been impressive events. Accounts of fireballs can be found in chronicles written in various parts of the world over the centuries. One of these rare events took place over the skies of Havana early in the evening of May 10, 1886, awing some, startling others, and terrifying many more. The event occured at 7:30 PM local time, as night was gathering over the city. Countless citizens watched as a large bright body moved slowly over the dome of the sky from a northerly direction.

At first the object appeared fixed in the sky, then it slowly appeared to pick up speed and grow larger in diameter and brightness as it moved southward, reaching a magnitude almost equalling that of a full moon, at which point it disintegrated into numerous fragments. Many thought that the event had concluded with this spectacular breakup and the glowing trails left by the fragments, but then its most impressive aspect began: a succesion of loud detonations, lasting for nearly a minute, shook the city. One should note that the darkening skies of early evening, coupled with the clear sky conditions, made the fireball over Havana appear all the more impressive.

The commotion caused by this sudden and extraordinary event forced the city's newspapers to seek an explanation for it, if for no other reason than to reassure a worried public, and to satisfy the curiosity of those wanting an explanation. Of course, there was only one person that could instruct and reassure one and all: the wise Father Benito Viñes.

Viñes was indoors when the event began, so he was only able to observe its final part. He calculated that the "explosion" of the meteorite, as he termed it, had occured some 15 miles from his location at the Observatory by taking into account the time lag between the observed breakup of the fireball and the arrival of the sound waves, as well as the speed of sound in air.

In reality, the sonic booms that accompanied the bolide were caused by the shock waves it created as it moved at supersonic speed though the atmosphere, and not because of any "explosion" of this extraterrestial body. One must, of course, understand that supersonic shock waves only began to be studied in ballistic tests conducted during the First World War. In the late nineteenth century, astronomers believed that bolides "exploded" in the atmosphere.

Viñes wrote a note that appered in the morning edition of the *Diario de la Marina* describing the phenomenon and the conclusions he drew from the observations:

THE AEROLITE OF THE NIGHT OF THE 10TH
The Observatory of the Royal College of Belén

May 12, 1886

The night before last, at approximately seven thirty, a meteor of extraordinary brilliance and magnitude was observed, followed by distant and prolongued rolling thunder. This brilliant meteor, which flooded the capital with bright light over the course of a few seconds and caused much alarm and anxiety, was in my opinion nothing but a notable aerolite that exploded a few leagues fron Havana in the lower layers of the atmosphere, many of whose scattered fragments may have fallen on the island.

In putting together my observations with that of some eyewitnesses, I deduce that it first appeared toward the NNW at a height of approximately 50 degrees above the horizon. At that point the meteor appeared as a small incandescent globe of constantly-increasing brilliance and an apparent diameter of 2 to 3 centimeters.

It passed a little to the east of our zenith as its apparent diameter was increasing rapidly and its brilliance was becoming dazzling, leaving behind a trail of very bright light. As it moved over the city, its nucleus was a bright red and its long tail was bluish white in color, somewhat resembling sparkling fireworks, though greater in brilliance.

As the meteor reached a position to the ESE of Havana at an altitude of 20 to 25 degrees above the horizon, the nucleus exploded in a multitude of bright sparks that flew off in different directions, followed by a prolonged thunder that lasted approximately one minute whose source appeared to shift toward the SE.

The large variations in the brilliance and the apparent diameter of the meteor indicate that it reached close to the surface before it exploded, and the loudness of the explosion at a distance of roughly six leagues is evidence that it occured in the relatively dense air of the lower atmosphere. In view of these facts, it would not be surprising to find that several fragments from the explosion might have landed somwehere on the island. —B. Viñes, S.J.

Other Aspects of Interest

The works of Benito Viñes were translated into English, French, and German and published by various editors abroad, both in full text and as abstracts. His work *Apuntes relativos a los huracanes de las Antillas* was published in English as *Practical Hints in Regard to West Indian Hurricanes* in three editions between 1885 and 1887. In 1889 Viñes' "laws" began appearing in the well-known "Pilot Charts" published by the U.S. Hydrographic Office[47] and were mentioned in the memoirs of Mr. Everett Hyden, who had spent

---

[47] The United States Hydrographic Office prepared and published maps, charts, and nautical books required in navigation. The office was established by an act of June 21, 1866 as part of the Bureau of Navigation, Department of the Navy. It was transferred to the Department of Defense on August 10, 1949. The office was abolished on July 10, 1962, replaced by the Naval Oceanographic Office.

some time in Havana visiting the Observatory the prior year. Even before, the American magazine *The Popular Science Monthly* had published in their July 1886 issue a summary of the monograph *The Storms of Cuba* (described above under item 11).

Viñes' *Investigaciones Relativas a la Circulación y Traslación Ciclónica* was translated into English as *Investigations of the Cyclonic Circulation and Translatory Movement of West Indian Hurricanes* by Eduardo Finlay. Two summaries of the same work appeared in German: one under the title *Neuerer Forschungen Uber West Indischen Orkane* and another in the weekly publication "Wochenschrift," which was published in the city of Halle, Germany (Gutiérrez-Lanza 1904).

It was obvious by that time that the high quality of several of Viñes' publications, based on his diligent work and well-thought-out ideas was being increasingly recognized by leading meteorologists around the world. The prestigious British publication Nature, published in London, and the well-known Les Mondes, edited in Paris by l'Abbe Moignó, both made references to the work of the meteorologist-priest and of the practical applications of his work. Viñes' work and his theories were mentioned in oral presentations at numerous European and American scientific meetings. We know of a presentation by Robert H. Sealt at the Royal Meteorological Society in London as an example of these presentations.

Belén Observatory's yearly data bulletins, which became more comprehensive over time under Viñes' direction, received a variety of accolades and prizes at international expositions. During the late nineteenth century, these events often served as a venue for scientific meetings. Among the awards received by these publications were a Diploma of Honor at the 1876 Centennial Exposition in Philadelphia, a Diploma and Silver Medal at the 1888 Paris Exposition, and a Diploma and Gold Medal at the 1888 Universal Exposition of Barcelona.

Before concluding this section, I must point out the valuable collection of press notes, containing forecasts and warnings that were issued when there was a threat of hurricanes. As mentioned before, these press clippings were collected at the Observatory in large albums. To keep these albums updated, workers at the Observatory had the ongoing task of scanning and filing relevant articles from the newspapers that were received several times a day.

According to Gutiérrez-Lanza, the Observatory kept similar albums containing a smaller volume of press clippings of items related to Cuban seismology or geology. These have been lost, unfortunately, making Viñes' contributions in these topics harder to assess.

By thoroughly analyzing the entire body of Viñes' work that has survived so far, I became more aware of something that he himself did not discuss or preserve for the future: his style of work. Using scattered notes to try to reconstruct what would have been his typical workday, I found that Viñes strictly adhered to a demanding work schedule. He was careful in his judgments, capable of applying rigorous scientific criteria in the solution of problems without resorting to shortcuts. This work ethic, coupled with his keen powers of observations, shaped his life as a scientist.

From the summaries of his notes that have survived it appears Viñes used the morning hours to analyze the observations he had recently made, as well as the cables that had been received overnight. By one or two in the afternoon Viñes had reached a conclusion about the meteorological situation of the day.

In one of Viñes' album of press clippings he added on the margin a brief note he sent to the periodical *El Pensamiento Español* (Spanish Thought). It appears that early in the morning of August 27, 1885 the editors of that publication requested an immediate weather forecast from Father Viñes. He responded in no uncertain terms that would not be possible, and explained his reason thus:

> *This paper goes to press at noon. When they brought me the request I told them that I could not commit to giving them my forecast by 12 because in order for me to do that I would have*

had to arrive at my forecast by 8 or 9, which is a time when I
am making observations and formulating my plan.

We also know that the first thing Viñes did upon waking early in the morning was to scan the skies, carefully noting all the characteristics of the clouds he saw, including their appearance and direction of movement. Next he would read the instruments and inspect those that had recording devices for later analysis, the meteograph in particular, and afterward he would read the newspapers that had been received at the Observatory, the cables, and any other information that might have been received at the nearby Navy headquarters and was regularly forwarded to him. Only at that point would he be prepared to issue his forecasts or other press releases.

Most of the press releases Viñes prepared were published in the evening papers or the following day's morning papers. If, however, there was a need to get the latest news out as soon as possible, they were published in special editions or as single sheet announcements that were printed and distributed as late as 8 PM. In times of threatening weather these special announcements were eagerly snapped up by the concerned citizenry.

With all these daily tasks, Viñes found very little time to sleep. The day was consumed by his Observatory routine and his liturgical obligations as a priest, among which included celebrating daily mass. At night, until late, he would read, write, and analyze data, noting important details that emerged that day. His eyes, relying on his small circular spectacles, must have suffered tremendous strain with the daily effort and the long hours of nighttime work by lamplight. His lack of sufficient rest and relentless work must have contributed to the progressive decline of his health during these years.

Viñes' contribution to meteorology was observational in nature, particularly the study of clouds. His principal achievement was developing a method of extracting the most information possible out of cloud-derived wind direction at different levels of the atmosphere under different weather patterns, particularly in the general vicinity of hurricanes. He studied in detail the typical cloud structure of hurricanes and used the knowledge he gained to deduce its probable future trajectory and evolution, often in terms of hours or days.

Viñes' statistical analyses allowed him to determine the average hurricane tracks during different months of the season in the Caribbean and Gulf of Mexico, as well as the mean latitude where their track recurvature took place. These data served as the basis for his later work forecasting hurricanes.

In his cloud work, Viñes used the cloud nomenclature proposed by the Englishman Luke Howard (1772-1843) in his 1803 work *On the Modification of Clouds*, although we do not know if he actually read that work. Already in his first year in Cuba, 1870, and his monograph on the Matanzas hurricane, his interest in clouds as precursor of hurricanes is evident.

By the same token, in his work *Apuntes Relativos* he dedicates more than 20 pages to observations of cloud movements and the conclusions that can be drawn from them, especially from cirriform clouds like cirrostratus and cirrus uncinus. As an example:

> *The cirrus and cirrostratus that precede a hurricane are always oriented in such a way that they emanate from the horizon and move toward the zenith. That direction can give us precisely, or to a good approximation, the location toward which the vortex of the hurricane is located, as if these clouds form an extension of the body of the hurricane itself...*
>
> *These cirrostratus serve therefore as banners that mark for us the speed and direction of the upper level currents of the hurricane.*
>
> *Through the orientation and outward movement of the cirrostratus from the radius of the cyclone one can see that the upper currents are divergent and centrifugal and that the center or focus of such divergence is precisely the vortex of the cyclone.*

Being that the surface currents of the hurricane are con-
vergent and centripetal, although modified by the rotation of
the Earth, as we have seen before, and that the upper currents
are divergent and centrifugal as demonstrated by the cirrus,
it inevitably follows, in my opinion, that the hurricane cirro-
stratus are due to the solidification of a contingent of vapors
carried unceasingly by the surface currents toward the axis
of the cyclone and from there ascending to the higher regions
where they spill out in divergent threads by virtue of the cen-
trifugal force developed in the rotation and counteracted at
such heights by lateral pressures. (Viñes 1877, p. 149)

The *Pilot Charts* published by the Hydrographic Office of
the United States acknowledged in 1889 that, guided only by the
movement of clouds, Viñes had been able to determine the track of
a hurricane from a distance of 900 miles. This constituted a huge
achievement for the time and gave him worldwide recognition, at
least among meteorologists and among those who, thanks to his
forecasts, were saved from perishing from the actions of air and
water either on land or at sea.

His Life and Times in Cuba

4

...In both the cities and the countryside of this island...I have always been surrounded by friendly people that have welcomed me warmly with the utmost hospitality. (Viñes 1877, p. 112)

Havana has always played a prominent role in the Americas. Its geographical location in Cuba, at the edge of the tropics near major shipping lanes, and marked by a gentle but occasionally violent climate, has made this island a strategically important place in the Western Hemisphere. Countless travelers, explorers, and geographers have been fascinated by the city. The pleasant sea breezes and gentle sounds of the breaking surf are occasionally interrupted by the roar of the hurricane, providing fascinating contrasts for locals and foreigners alike.

One of the earliest Cuban historians, José Martín Félix de Arrate (1701-1764), in his work *Key to the New World, Gateway to the West Indies,* published in the late eighteenth century, refers to Havana this way:

Havana was founded at twenty three degrees, ten minutes north, though Herrera puts it one minute further south; its nature is warm and dry like the rest of the island, its skies clear and cheerful, because the winds that generally prevail along the coast unburden the sky of clouds, making the heat more bearable and the storms that prevail from June to August less fearsome and damaging, though they be accompanied by lightning. Such is the price paid for such benign climate. (Arrate 1964, p. 75)

It was in Havana, this long celebrated city, where Viñes lived and worked during the most productive years of his life; this is where he dedicated his life to the study of the earth sciences, to his religious order, and, serving as a sentinel of impending natural disasters, to the service of society at large. From the Observatory windows, he could survey the splendid bay and its surrounding hills. He left the city several times in pursuit of various objectives, but he always returned, and he remains here to this day, still among us.

Much has been written about the influence that weather has on our senses. Viñes was no exception; while he lived in Havana he must have found stimulation in things like the bracing cool air from the thunderstorm outflow, the loud roar of the hurricane winds, and the cold sensations of raindrops falling on his skin. Viñes' writings attest to this sensibility, as they did to his early appreciation of the important role that meteorologists can play in society.

Viñes planned for and tried to implement a network of weather stations at strategic outposts that would allow him enough lead time to prepare more timely and accurate forecasts. To secure the needed resources, he had to struggle against obstacles posed by widespread ignorance about the nature of tropical storms and by societal and budgetary priorities that were placed elsewhere.

He was not only concerned about gathering the data he needed for the preparation of his forecasts, but also about how the forecasts would be used and disseminated, both in Cuba and other parts of the Caribbean. His priority was to obtain meteorological data from the eastern Caribbean, including the island arc of the

Lesser Antilles, a region that often experiences hurricanes before they hit Cuba.

Diario de la Marina, in its September 26, 1876 edition, published an editorial that included the following lines:

> *We think it would be very convenient for this interesting study to tap the resources of amateur weather observers in Santiago de Cuba, Cienfuegos and other cities of the island, without the need to spend public funds... We already have an established observatory at the College of Belén, led by a very accomplished person. When we are faced with similar threats in the future, if all these observers were to be coordinated in advance from a central location and if they had free use of a telegraph so they could report in a timely way whatever weather observations they make to a single coordinating center... Without coordination fewer benefits would accrue...*
>
> *We suggest that the appropriate personnel, especially His Excellency the Commandant of the Navy, give due consideration to this idea and issue the appropriate decrees to implement it.*

The writer, in another section of the article, uses the recent occurrence of a hurricane to remind the government that a few years before funds had been allocated for building and equipping an observatory in the "Professional Schools" (referring to the aforementioned Physical-Meteorological Observatory of Havana) without any positive results resulting from that investment. With this editorial, the newspaper began a lobbying effort for the establishment of a network of weather stations.

The newspaper not only proposed the use of telegraph—the leading-edge technology of the time—as the basic communication tool for the planned weather station network, but also supported the idea of creating a central clearinghouse at the Belén Observatory where all the data would be received and analyzed. It also indirectly suggested that the telegraph companies should provide the service to support the weather station network as a free public service.

Just two days later, Viñes laid out his proposal under the title "A Useful Project" in the daily *La Voz de Cuba*. As he also explained in his book *Apuntes Relativos a los Huracanes de las Antillas…*. Viñes proposed:

> *In the case of hurricanes, it is of utmost importance to acquire data in as many locations as possible …I have insisted on other occasions, and will continue to insist on the importance of these data. It is necessary for all to realize that without the availability of simultaneous observations made at different locations it is not possible to solve the problems that arise in each particular case.* (Viñes 1877)

Unfortunately, it took another ten years for this project to materialize. Cuba's colonial administration was using every resource at hand, both political and economic, to fight the ongoing armed insurrection pursuing the goal of independence for Cuba and to prevent the rebellion from spilling out of the eastern provinces, where it had been contained after eight long years of a bloody and costly war. The colonial government had little interest in allocating funds for weather research; in spite of knowing full well that the Spanish Navy continued to be at the mercy of tropical storms during the hurricane season, year after year.

It is precisely because of these circumstances that the Navy collaborated informally with Viñes, sending him all the ship weather observations it could muster, especially during hurricane season. The nearby Navy headquarters forwarded weather reports to Viñes from as soon as they were received. He, in turn, gave the Navy highest priority in his forecasts. We have already mentioned a weather report sent to Viñes by the Navy in 1875; several dozen more examples could easily be cited.

Below is a report, typical of those Viñes sent to the Navy Headquarters to alert them about a potentially dangerous situation:

20th August 1879 at 9 A.M.

Don Luis García Carbonell
Secretary to his Excellency the Commandant General
of the Navy

Dear Sir:

There are indications of a gyratory storm in the third quadrant; it is possible this time of year that such a storm will move toward the Gulf of Mexico by way of the Yucatan Channel or the mainland of Yucatan toward the coast of Texas. It is probable that the effects of the storm will be felt in the western tip of the island.

I would be grateful if you could communicate my assessment to the Harbormaster, whom I had the pleasure of seeing a few days ago.

Your humble servant,
Benito Viñes, S.J.

Inspection of the albums of press clipping found in the College archives reveals that this collaboration became closer from 1879 on, always as an informal and mutually useful exchange of information, especially during the hurricane season.

This relationship was eventually formalized in 1883 with an ordinance issued by the colonial government on September 21, which directed that all shipboard weather observations should be forwarded to the Belén Observatory as soon as the ships arrived at Cuban ports. For years afterwards that was the best arrangement Viñes could achieve.

In July 1880, the Cuban press once again took up the idea of creating a network of meteorological stations, this time coming from the *El Bien Público* newspaper, published in Santiago de Cuba.

We are already in the second half of July; the autumnal equinox is approaching, the season when gyratory storms often visit us, causing considerable damage to cities and countryside both, and we have yet to hear the fate of the proposal made by Fr. Viñas [sic]... we hope that the newspapers of the capital, which are more likely to have the ear of the authorities, call to their attention the merits of organizing, during the storm season, a system of signals throughout the island that would revive the project of Fr. Viñas [sic], an idea we would like to see materialize. (El Bien Público 1880)

As one can easily imagine, this call fell on deaf ears that did not begin listening until September and October of 1886, a year in which three relatively weak storms accompanied by heavy rains damaged large portions of western Cuba. Already by the July 18 of that year, *La Voz de Cuba* was once again calling for the provision of funds for the establishment of the observation network. This time, however, Viñes himself wrote the newspaper column, in which he exhorted those whose economic interests were at stake to help fund the establishment of the network, arguing that that with a small investment much greater losses could be avoided.

Although not expressing it directly, Viñes used this article to criticize the shortsightedness and greed of those who were always happy to make use of his forecasts but continued to ignore his pleas for modest private contributions to fund his proposal.

Why wouldn't our steam ship companies, insurance companies, and others engaged in commerce contribute to the meteorological service in times of hurricanes, allocating small amounts of money in order to receive telegrams from wind-

ward locations at critical times? With those funds, added to others received from other parts of the island, we would not be caught unawares of the arrival of a hurricane and they could protect their interests at minimal cost. (La Voz de Cuba 1886)

Later in this article, Viñes summarizes his argument by writing "with these funds I would be able to better determine the march of cyclones and expand the radius of coverage." The serious damages caused by the storms of 1886 that followed the publication of this article added urgency to his plea.

Confronted by this situation, and aided by the relative peace that reigned in the island at the time, the authorities realized they had to do something about this matter. This time the loudest clamor came from the largest players in the private sector. Finally, on September 7, 1886, even as a cyclone was crossing the Caribbean just south of the westernmost province of Pinar del Rio, the Havana Board of Commerce called a special meeting, where one of its leading members, Mr. Narciso Gelats, presented a proposal for the Board to devote resources to the goal of setting up a telegraph notification system "related to atmospheric variations." The Board's resolution was published by the Havana press. An item from *La Iberia* reads:

LA IBERIA, *October 12 1886*
IMPORTANT RESOLUTION

The Havana Board of Commerce met...and approved the motion made by Mr. Gelats to provide resources to the distinguished meteorologist Rev. Fr. Viñes, Director of the Observatory of the College of Belén... This project is of undeniable utility, and of course we congratulate the Board on its action and hope that this resolution be implemented promptly, so that our distinguished and wise friend, the Rev. Fr. Viñes have at his disposal all the resources he needs to give us timely warnings... (La Iberia 1886)

It is not surprising to note that this news was received at the Observatory with much skepticism. After so many failures and false starts, skepticism became the default assumption made by these dedicated men. Somebody, perhaps Father Alberdi or another religious assigned to the task of compiling press clippings wrote in pencil below this article the following note: SURE, GO AND TIE THESE FLIES BY THEIR TAILS[48]... a sarcastic comment written with biting humor, given all the previous promises that had been made and not carried out. Whoever wrote that was not far off the mark.

Regardless, since this constituted a landmark event—the establishment of the first meteorological observational network in the Caribbean—it is worth a closer look. The Board, likely at the behest of Viñes himself, selected seven locations in the Caribbean basin to form part of the network, all located east of Havana. The operators of the first four stations, in the islands of Trinidad, Martinique, Antigua, and the city of Mayagüez in Puerto Rico were to make observations and submit them to the Maritime Insurance Company of New York, coordinated by Mr. Aquilino Ordóñez, a member of the New York Board of Underwriters.

The other three stations—if one could apply that name to them—were located in Barbados, Jamaica and Santiago de Cuba, their observers paid by Her Britannic Majesty's consul in Santiago, Mr. Frederick W. Ramsen.

Transmissions of the observational data were made with the use of a code, which used less "bandwidth" and consequently saved money. The agreed-to code went as follows: 1) barometric pressure; 2) wind direction; 3) wind speed expressed in numbers from 0 to 4: where 0 indicates calm, 1 ordinary breeze, 2 strong breeze, 3 gusts, and 4 hurricane; 4) direction of movement of low clouds, and 5) characteristics of cirrus clouds, including their direction of movement, etc.

---

[48] An archaic expression roughly meaning "I'll believe it when I see it."

Once details of the network were worked out, in accordance with the resolution it had adopted, the Board formally notified Viñes and the College. This was done by a delegation comprised of Mr. Gelats and nine other members, who visited the College of Belén on October 13, 1886. They were received by the Rector (President) of the College, Father Isidoro Zameza, and by Father Viñes, Director of the Observatory, who took formal receipt of the following document:

DIRECTORATE OF THE GENERAL BOARD
OF COMMERCE OF HAVANA

The Executive Committee of this Board, in its session of the 7th of the past month, after having discussed the valuable services that you have been providing to commerce and navigation by making announcements of atmospheric perturbations by dint of the perseverance and zeal with which you have been making meteorological observations, a service which greatly merits encouragement, resolved that a commission of its members should convey to you the high regard in which you are held by all classes of society, and particularly by those most concerned about the march of cyclones that so frequently punish these latitudes. The Executive Committee also directed said commission to coordinate with you the most expedient manner of providing you with telegraphic notices that would facilitate your important observations, to which end, and in order to notify you of this resolution, said commission will have the honor of personally conveying this document to you.

May God guard you for many years.

Havana, October 1st 1886
Narciso Gelats to Rev. Fr. Benito Viñes

To this document—with its stilted language—which included a vague offer of financial help, Viñes had a formal response, from which we cull the following quote:

…The laudable resolution of the General Board of Commerce will represent for me a most valuable help with which I can continue my investigations with greater likelihood of success. I am the first to bless with all my soul such philanthropic determination…

Havana, October 14th of 1886
Benito Viñes S. J. to Mr. Narciso Gelats, President, etc.
(Gutiérrez-Lanza 1904, p. 81).

All was left at this point was to work out the details. The contract for the cable transmission was negotiated with two telegraph companies named Cuba Submarine and West Indian and Panama. Mr. Juan de Musset conducted the negotiations on behalf of the Board of Commerce, while Viñes himself represented the Observatory. Also participating were Mr. Eugenio Fortún of the Cable Commission and Mr. Bernardo Arrondo, Controller of Communications.

The accord that was reached gave the Observatory the advantage of paying for the cables in Havana at a 50% discount, in addition to the exclusive use of the aforementioned codes that reduced the costs by condensing the volume of transmissions. Cables sent abroad from Cuba did not incur charges.

Viñes requested two transmissions per day from observers: at 7 AM and 3 PM. The service was formally inaugurated on September 10, 1887. The Board of Commerce was dissolved the following year, 1888, to be replaced by the Official Chamber of Commerce, Industry and Navigation, which continued maintaining, though not always reliably, the communications network established by the former Board of Commerce during the months with the greatest threat of tropical storms: August, September, and October during the years 1888 and 1889.

That was all that could be accomplished with the observational network Viñes had worked so hard to implement. We cannot be sure how efficient and reliable this service was, but Viñes continued receiving information from these locations, and from other locations outside the network through informal exchanges for years before and after, some mentioned in his communiqués as early as 1877.

Financial records from that time kept at the College show that the deficit incurred in keeping the "stations" from Jamaica, Barbados, Saint Thomas, and Jamaica grew from $125.00 in 1891 to $160.00 in 1892. Even during the tenure of Father Gangoiti, Viñes' successor as Director of the Observatory, the deficit had not been fully paid off, so if the network participants were still making observations, they were either volunteering their services or receiving payments Ramsen received from other sources.

In spite of these setbacks, Fr. Viñes-by himself or with the help of others-continued his demanding work, something that earned him a high level of trust among Cubans, particularly the residents of Havana, and an international reputation as well. The reservoir of good will created by Father Viñes remained with his Jesuit institution for more than seven decades afterwards.

In the newspapers of the time one could frequently find praise for his accurate forecasts, and plenty of evidence to show how closely his forecasts were followed. Witness this note that appeared in the Boletín Mercantil (Mercantile Bulletin) of Puerto Rico, showing the level of confidence in the Observatory of the College of Belén and its director in the neighbor island of Puerto Rico. This note is dated August 1, 1880 during the hurricane season.

We are in the dark… there are rumors floating around that the wise Father Viñes, Director of the Belén Observatory, is predicting that a cyclone will form north of the Island. We believe that this is misinformation concocted by those who don't know the facts in order to alarm pusillanimous people. We that do not indulge in these pursuits can reassure the reader that no such news has been published in Havana newspapers… So those that are concerned can rest assured. (Boletín Mercantil 1880)

Widespread opinion, not shared by Fr. Gutiérrez-Lanza, claims Viñes always refused direct help from either the Spanish or American government, the latter of which was supposedly was trying to absorb the Belén Observatory into the *Signal Service*.

It would not be surprising to learn that—in spite of the Observatory's financial difficulties—Viñes would be reluctant to enter into any arrangements with the government that would compromise the independence of the Observatory or the College. Suffice it to say that the College of Belén had paid for 30% of the cost of the instruments Viñes acquired in Europe as well as the import duties that were due upon arrival in the island, the last being evidence of the lack of support for his enterprise shown by the colonial government's refusal to waive these tariffs.

The Observatory received requests, almost on a daily basis, from individuals planning to travel abroad, sending or receiving valuable mail or cargo, or just seeking assurance of good weather for an uneventful voyage. On many occasions, before sailing from ports such as San Juan, Galveston, Tampa, Santo Domingo or New York, ship captains would check in with Viñes for his latest forecasts.

Cable traffic was particularly intense between Havana and cities in the Caribbean and the southern United States between the months of August and October, the height of the Atlantic and Caribbean hurricane season. One could frequently find notes such as this one in the newspapers of the day:

The Rev. Father de Corrieres has the pleasure of announcing to the students that will accompany him to Spring Hill that the steamship Borrusia will be departing tomorrow. According to a report by Rev. Father Viñes, the voyage will occur under safe (weather) conditions.... (La Voz de Cuba 1879)

Parents sending their children abroad would of course be reassured when Viñes confirmed that no hurricanes were expected to interfere with the trip. Another example is worth citing. This one involved the city fathers of Batabanó, a coastal town on the south shore of the island near Havana, who, when faced with a sudden change in the weather, sent a telegram to the Governor's office, which forwarded a copy of the following telegram to the Belén Observatory:

Mayor of Batabanó at 7:00 PM. To the Governor. Havana. At 2:30 this afternoon a suddenly strong wind began blowing from the SE, causing damage to ships in port. The winds persisted and got stronger by 6:30 PM. I beseech Your Excellency to inquire with Fr. Viñes if there is a threat to this location so that we can take the proper precautions. I shall await an answer at the telegraph station.

Mayor Bustillo.

One can clearly sense the alarm that must have prevailed in that town that led the mayor to remain at the telegraph office until he received a reply from the Observatory. Once told that the winds were a result of a particularly strong "sur" (south wind event), and not a cyclonic disturbance, the mayor was reassured, and through him the town's citizens as well.

Let us now turn to some aspects of Viñes' relations with the newspapers that represented various sectors of the political spectrum, and how he was generally viewed by the print media.

Weather reports were routinely sent as "scoops" to two newspapers: *Diario de la Marina* and *La Voz de Cuba*. Other newspapers reprinted these reports in later editions. One can see how these two newspapers would be given preferential treatment, since they catered primarily to a Spanish-born and Catholic readership.

Another daily that received meteorological information from Belén was *El Triunfo*. That Havana newspaper, with a liberal and anticlerical editorial bent, irreparably damaged the working relationship it had previously enjoyed with Viñes in 1880. It published an article in which the newspaper questioned the integrity of the Jesuit order by reprinting an unfounded rumor that claimed the Jesuits had stolen a fortune in Brazil originally belonging to King John VI of Portugal. According to the rumor, the stolen goods were found years later when a convent belonging to the order was demolished.

Viñes reacted to this slanderous "news item," as one might expect, by suspending delivery of his weather reports to *El Triunfo*. These events began a drawn-out controversy between that newspaper and the Belén Observatory that was never fully healed.

Added to this incident was another that had been festering for a while: *El Triunfo* had published several articles written by the French philosopher Ernest Renan (1823-1892) that questioned the divinity of Jesus, and timed those articles to coincide with major Christian holidays such as Christmas and Easter. Those provocations would obviously not have passed unnoticed at Belén College.

Things came to a head in August and September of 1880 when Viñes permanently cut off delivery of information to *El Triunfo*. The editors of the paper retaliated by blaming Viñes for this breach by calling him "miserly." Given that *ad hominem* attack, Viñes had no recourse but to defend himself by having the *Diario de la Marina* publish a letter he had written to the editors of *El Triunfo*. It read thus:

To the Editor of El Triunfo

Dear Sir

In the edition of El Triunfo published on Saturday September 21st, I had the occasion to read the following lines:

"Our readers should not think that the reason for not publishing the reports from the illustrious Father Viñes, even as they appeared in other newspapers, was our oversight. Yesterday, as was our custom, we went to get the report and we were told that they were not permitted to share this information with us. It is lamentable that the Rector of Belén College is not aware of the inconvenience of depriving the public of knowledge of approaching hurricanes. How good are scientific advances if their benefits are not widely shared?"

My refusal to share this information, which appears under my name in other newspapers, was in accordance with the wishes of the Reverend Father Rector of this College. I don't see how that could be viewed as depriving the public of awareness of approaching storms, as you state, since my weather reports appear in newspapers in this capital such as: "El Diario de la Marina", "La Voz de Cuba", "El Boletín Comercial" and "La Correspondencia". In addition, "El Triunfo" should understand that our refusal to share information is due to their often repeated declarations that are contrary to the Roman Catholic faith that is practiced in Cuba and under whose banner the Company of Jesus plays an active role, and in whose bosom we are fortunate to live and die.

In the hope that you might see it fit to print this letter, I remain your humble servant.

Benito Viñes S.J. (Diario de la Marina 1880)

El Triunfo apparently suffered a significant drop in circulation (and revenues) as a result of Viñes' boycott, since his weather reports were in high demand by its readers. Faced with this situation, *El Triunfo* resorted to its last-resort weapon: ridiculing Viñes with the purpose of undermining his credibility. On August 28, 1880, the newspaper published this sarcastic note:

TWENTIETH IRON[49]

The meteorologist who wrote a letter to Mr. M. advising him not to board the "Niagara" because he would encounter a cyclone in the voyage, dropped another iron with the officious notice he received, since according to him, the voyage took place in excellent weather.

These efforts were all in vain. In spite of its protestations, *El Triunfo* never got Viñes' weather reports again.

A year later, a similar situation occurred with another Havana newspaper, this one called *La Discusión* (The Discussion), which, in solidarity with *El Triunfo*, launched a rhetorical attack on the Observatory in the person of Viñes. On September 1, 1881, it published the following anonymous epigram, which rhymed in the original Spanish. It appeared without a title:

Dialogue

Will a cyclone happen, Jose?
–Well, I cannot understand it,
nor can I now track its movement
but I don't know its trajectory
all this is due to my honesty
so, no, I don't know where it's going!
–Won't they let you talk at home?
Friend, let me tell you the truth

---

[49] The upper case title "VIGESIMA PLANCHA" as it appeared in the original Spanish text can be translated roughly as "TWENTIETH IRON". The colloquial expression "dropping an iron" can also be interpreted as "the ridicule emanating from a mistake one makes", presumably because the loud noise made when the solid metal implements then used in ironing clothes were accidentally dropped on the floor, drawing attention to the event from people in the vicinity.

that they all say that you're so skillful
that they all say you're resourceful
but I witness with much pain
that your lips quite sealed remain
–Is that what the people say?
–They say that, I'm quite serious,
that day and night you're oblivious
following the meteors
–Well, they are correct 'cause it's right
devoting my life for mankind
these count as my most precious treasures:
to lend my skilled hand to my brothers
to teach them the goals of my science
and all that my experience covers
as a good true Catholic should
–And the rest?
–Well they do not now deserve
to receive our calculations
nor from our fountains imbibe
–Holy Christ, what if they perish
in the agitated sea?
–Well I will not change my behavior
obeying is what should strive for
so let them now learn how to swim!

The "José" in these lines, of course, refers to Viñes, one of whose middle names was in fact José. As if this were not enough, the same article satirized the arguments made by Viñes in his earlier letter to the editor of *El Triunfo* months earlier. These acrimonious exchanges seem to have been the only disputes in which Viñes was involved.

As a result of these events, Belén was forced to adopt a policy with respect to the press outlets that requested Viñes' weather forecasts: first, that they should accord proper respect to the Catholic religion, its dogmas, and its priests; and second, that the College, the Observatory, and their respective directors should be treated respectfully by their writers.

Viñes, like other meteorologists to this day, did not escape being lampooned as a result of a failed forecast. On October 11, 1886, *La Iberia* published the following note:

> *...It was being said on the street yesterday that the cancellation of the bullfight was due to Fr. Viñes' forecast. We beseech the wise Jesuit not to issue forecasts in advance of bullfights, to spare us disappointments like that we experienced yesterday afternoon, and that way keep fans from speaking ill of the Society to which he belongs....* (La Iberia 1886)

In response to this sarcastic item, Viñes sent word to this newspaper to cancel the Observatory's subscription and notified them that the delivery of his forecasts would cease. The editors, seeing the threat this would pose to their circulation, quickly backtracked. Eleven days later, on October 22, it published a note praising Viñes and the Observatory: *"Illustrious Jesuit... respected and blessed by all... etc., etc."* *La Iberia* could read the handwriting on the wall and did not want to repeat the negative experiences of *El Triunfo* and *La Discusión*.

There is an interesting handwritten note by Fr. Alberdi on the margin of the album page containing this press clipping. Bracketed by exclamation points[50] and underlined, it reads: *"To no avail, they tried here and there through intermediaries to get us to resume communications, but didn't succeed."*

---

[50] In Spanish usage, an "upside down" exclamation point appears at the beginning of the sentence in addition to the one at the end.

Viñes' good nature eventually won out. *La Iberia*'s apology was accepted and the delivery of forecasts to them resumed. However, the practice of making fun of meteorologists after failed forecasts will apparently be with us forever.

Notwithstanding the aforementioned events, the Havana print media generally maintained good collaborations with the Observatory, as there was no doubt of the intense interest the population had in their forecasts and the anxiety with which they were awaited when hurricanes threatened.

Apart from these matters, which show how Viñes could be, in turn, delicate, demanding, emphatic, and forgiving, newspapers of all stripes always held him in high esteem as a scientist and as a priest.

The Belén Observatory, under the leadership of Benito Viñes, maintained fairly close relations with the United States Signal Service, which would eventually become the Weather Bureau and is now the National Weather Service. Belén and Washington began collaborating in 1877. Two years later, the Observatory started making use of meteorological observations that the Signal Service observers were reporting from Caribbean locations. In return, Viñes would share with the Americans his hurricane forecasts for the area (Viñes 1877, p. 14).

Contacts between Viñes and the Signal Service likely became more frequent after his visit to the United States, during the 1870s, the exact date of which is still unknown. Viñes, however, in the prologue to the 1885 observation summary, mentioned this trip in passing. It was probably during his U.S. trip that foundations were laid for future collaborations between both institutions. Evidently, these collaborations were sustained and systematic, since copies of communiqués issued by Belén Observatory almost always were annotated with "sent to Washington."

Other documents in the archives of the Observatory confirm the close working relationship between both meteorological centers. There one can find two letters Viñes received from Adolphus

Greely,[51] polar explorer and distinguished meteorologist, who headed the Signal Service at the time. In these letters, Greely expresses his high esteem for Viñes' work on tropical cyclones. He states: *"your published memoirs about these storms have been read with great interest in this office"*.

Unfortunately, American law prohibited paying foreign meteorological observers, but Viñes' reports were forwarded to the Signal Service because the cable company charged the outgoing telegraph fees to Washington.

Evidence of the routine collaboration between the two institutions can be seen in a second letter from Greely in the Observatory archives:

SIGNAL OFFICE
WAR DEPARTMENT

Reverend Benito Viñes, S. J.
Royal College of Belén, Havana, Island of Cuba

Washington, August 31, 1887

Dear Father,

In acknowledging receipt of your letter of the 6th of this month, I also want to express my thanks for your telegrams informing this office of the march of cyclones.

---

[51] Adolphus Greely (1844–1935). Among his many achievements included exploring the northwest coast of Greenland, setting the "farthest north" record of the time at 83°23' 8" latitude as part of an expedition where most perished, serving as military commander in the emergency situation created by the 1906 San Francisco earthquake, and receiving the Congressional Medal of Honor.

It goes without saying how valuable we consider these telegrams, both to this office and to the maritime interests of the United States. At your request, I am enclosing six copies of the "Instructions for Sending Telegrams about Hurricanes," 1881 edition, and we request that you follow the code explained therein in the preparation of any future communications you kindly send to us.

Yours sincerely,

A. W. Greely
(Archives of the Cuban Institute of Meteorology, 1881–1885)

The relations between Belén and the Americans remained uninterrupted for almost three decades, until they were severed at the start of the Spanish American War[52] in 1898.

An event occurred in 1890 that revived hopes that an international observational network with headquarters in Havana could be established. That year, Cuba's colonial administration received a Royal Edict from Madrid that directed the establishment of an official observatory under government control that would produce high quality observations and conduct research. On April 9, 1890 the colonial government (officially The General Government of the Island of Cuba) officially requested that the Royal Academy of Medical, Physical and Natural Sciences of Havana, and the College of Belén each prepare a report on the feasibility of constructing a new observatory. The intention was to fill the gap created by the failure of the Physical-Meteorological Observatory of Havana that had occurred years before.

Evidently Viñes was the go-to man for this task. The Academy of Sciences sent his distinguished member the following missive signed by its secretary:

---

[52] Known in Cuba as the Spanish-Cuban-American War.

I have the honor of inviting you to accept your appointment as member of the Commission that the Academy has entrusted with preparing a report to the General Government about the establishment of a Meteorological Observatory in this city, and even though your position as Fellow of this Academy exempts you from this kind of task, the Academy hopes you will not withhold your valuable cooperation in a matter of such public interest.

May God keep you in good health.
Havana, April 19 /980 [sic].
(Archives of the Cuban Academy of Sciences, Benito Viñes S.J.'s file)

The Father, as could be expected, had eagerly accepted his charge, and after obtaining permission from the Rector he sent his reply a few days later:

Mr. José Torralbas
Secretary of the Academy of Sciences

I have the honor of acknowledging receipt of your letter of the 19th of this month, in which you kindly invite me to accept my nomination as member of the Commission charged with informing the General Government about the establishment of a meteorological observatory in this city. I gladly accept the invitation, and will be honored to form part of the Commission, if the Academy does not find it inconvenient to know that I must also prepare a separate report to the General Government about the same matter. Kindly communicate to the Academy my gratitude for the honor they bestow on me with this appointment.

May God guard you
Havana, 25 April 1890

The Commission was officially constituted on April 19, 1890 at the Belén Observatory. It was composed of Don Francisco Paradela, Don Adolfo Sáenz, Dr. Carlos Theye, and of course Father Viñes.

Both reports, that from the Academy and Viñes' own report from the Observatory agreed that the proposed observatory ought to be of "first order" and centralized, that is, with all the features that a national observatory should have, with satellite stations in Cuba and abroad. In addition, the reports called for the proposed observatory to help support the agricultural sector of the island (Annals of the Academy, V. 27, p. 172).

In Viñes' report it was evident that he harbored doubts as to whether this project would come to fruition once the operating costs were taken into account. He understood clearly the difficult financial situation faced by the Spanish Government at the time, as well as the unreliability of the Chamber of Commerce's support for the observational network.

The Commission report, however, suggested an alternative and less costly possibility: expanding and strengthening the existing network of stations in the country and other locations in the Caribbean, all of which sent the observations to Belén. Its report was delivered to the government on May 6, 1890 after being approved for release by the Academy. Viñes, who with typical modesty refused to accept an honorarium for his services, did not sign it.

The Commission's proposals were never implemented. The report's historical significance lies primarily in allowing us to evaluate the broad support and respect that Viñes had gained among his contemporaries, at high levels of the colonial government, as well as in the scientific community of the island.

The work of Benito Viñes transcended his time, since his ideas and theories were taken into account by meteorologists from many nations in the first half of the twentieth century. In the list of references of many texts dealing with tropical meteorology, starting in 1899 in particular, Simon Sarasola,[53] William Ferrell,[54] Ivan Tannehill,[55] and José Carlos Millás[56] were among the many that pushed forward the work that Viñes had begun.

Around the turn of the twentieth century, Viñes' ideas marked a watershed in the study of tropical meteorology. Several scientists, mainly from the United States, travelled to Havana during the years Viñes was at the helm of Belén to consult with him in matters related to hurricanes.

Everett Hayden[57], then serving as marine meteorologist at the U.S. Hydrographic Office and Editor of the monthly *Pilot Chart* publications, visited Cuba in 1888. Upon his return to the U.S. he gave a series of talks on the subject of tropical cyclones and even went as far as proposing to rename the term "hurricane" after Viñes!

---

[53] Simon Sarasola (1871-1947) came to Havana in 1897 after the death of Viñes, where he became acquainted with his work. He later worked at Georgetown University. After being ordained a priest in Maryland, he returned to Cuba to found a meteorological observatory in the city of Cienfuegos. He went on to establish the Meteorological Service of Colombia in 1921. He later returned to Spain, where he worked as a meteorologist at the Air Ministry until his death.

[54] William Ferrel (1817-1891), an American meteorologist, developed a theory to explain the mid-latitude atmospheric circulation cell in detail, and it is after him that the Ferrel cell is named.

[55] Ivan Ray Tannehill (1890-1959) was a lieutenant at Fort Story, Virginia soon after World War I, and later became a forecaster with the United States Weather Bureau and a prolific writer, focusing on meteorology. His text on hurricanes remained the defining work on the topic from the late 1930s into the early 1950s.

[56] José Carlos Millás (1889-1965) was a Cuban meteorologist. He is known for his research on past *Atlantic hurricane seasons*, and has been called one of the "fathers of tropical meteorology."

[57] Edward Everett Hayden (1858-1932) was a career Navy Officer that reached the rank of Rear Admiral. During the period 1887-1893 he served as marine meteorologist and editor of pilot charts at the Hydrographic Office. He is buried at Arlington National Cemetery.

It is also important to mention a little known aspect of Viñes' life that other biographers have not mentioned: the close friendship with fellow scientist Carlos Finlay[58], a medical doctor and one of Cuba's most distinguished scientists ever.

Their friendship began in the early 1870s and was marked by a deep sense of mutual loyalty. They shared several common interests: Finlay served as physician to the College; both men were members of the Academy of Sciences. It was a friendship between priest and layman, Spaniard and Cuban, which transcended the passions of place and time marked by a bitter and bloody war of independence.

Viñes was a frequent guest at Finlay's residence (López Sánchez, 1987) and among the products of that friendship was an ongoing exchange of views about science in general and their respective fields in particular. Dr. José Lopez Sanchez, Finlay's biographer, claims that Finlay acquired his knowledge of Physics and Meteorology through his interactions with Viñes.

The two friends must have had long conversations about Finlay's childhood experiences at *"Buena Esperanza,"* his family's estate and about two hurricanes that Finlay must have experienced as a child that were named after saints' feast days: the *"Storm of St. Francis of Assisi"* in 1844 (October 4th) and the *"Storm of St. Francis Borgia"* in 1846 (October 6th) that swept thought the province of Havana and damaged the coffee crops from which the Finlay family derived a portion of its income. On his part, Viñes must have kept Finlay apprised of topics such as the progress of

---

[58] Carlos Finlay (1833-1915) was a Cuban physician and scientist who was the first to theorize, in 1881, that a mosquito was a carrier of the organism causing yellow fever. His hypothesis and exhaustive proofs were confirmed nearly twenty years later by the Walter Reed Commission of 1900. Finlay went on to become the chief health officer of Cuba from 1902 to 1909. Although Dr. Reed received much of the credit in history books for "beating" yellow fever, Reed himself credited Dr. Finlay with the discovery of the yellow fever vector.

his hurricane research, his planned trips and the barriers he was encountering in setting up his observational network.

Viñes' academic file at Belén contains notes that accompanied copies of his *"Magnetic and Meteorological Observations at the College of Belén"* which were edited at the Observatory. Finlay's name frequently appears in the list of recipients of copies of these publications.

The first paper Finlay presented the Academy, assembled in formal session on 22 September 1872 was on a meteorological topic: *"Atmospheric Alkalinity Observed at Havana"* (Annals of the Royal Academy, 1872, p.183). He based a significant portion of his paper on meteorological data recorded in the city. The historical importance of this paper is that it was the first he presented at that institution, as part of his admission as a member of the Academy.

We still do not know if Viñes was instrumental in helping Finlay with the meteorological part of his paper, but we know that the latter continued his observations in May 1873, in particular on days 1-9 and 26-31 of that month. Volume 10 of the Academy Annals give details of the meteorological data Finlay used in his work. These included wind direction, air temperature, atmospheric pressure, and humidity. These data were obtained by direct reading of the instruments, all made in period between 3:30 and 4:00 PM. Finlay states in these notes:

> *Regarding the meteorological observations in this table, I must add they are very reliable, due to the fact that the alkalimetric[59] measurements were also made at the same location, the Observatory of Belén College; and I am taking this opportunity to express my gratitude to its distinguished Director, and Fellow of*

---

[59] Alkalimetry is the specialized analytic use of acid-base titration to determine the concentration of a basic (synonymous to alkaline) substance. It is an example of acid-base titration, the determination of the concentration of an acid or base by exactly neutralizing the acid or base with an acid or base of known concentration.

this Society for the kindness he has shown me by making these data available to me and also for the interest he has shown in my experiments, which were all but one conducted in his presence. (Annals of the Royal Academy, 1873, p. 45)

As an aside, six of the fifteen observations note the presence of rain or drizzle, and there were two observations with thunderstorms in progress.

Finlay faced difficulties presented by the enmity of an Academy colleague: Dr. Marcos Melero, which precipitated an event instigated by the latter, with the apparent intent of undermining Finlay's credibility. Here we find evidence of Viñes' intercession in defense of his friend; it is worth a small digression. In the words of Dr. López Sánchez:

During the public session of 22 June 1873, he (referring to Melero) *read a note where he imputes Finlay's work, arguing that not only are the experiments erroneous, but also that they lack scientific base; that the data he presents in support of his work are incomplete and biased. The debate turns into an acerbic discussion in which words were spoken in a disrespectful and violent tone by Melero. Finlay kept his equanimity, and asked Melero for any proof he had to substantiate his claim, of which Melero had none....*

Melero then pretended to refer to another work by Finlay, which he had misplaced some time before.

President Nicolás J Gutiérrez replied that this was a sub judice[60] *matter, which would require him to stop talking at*

---

[60] In law, **sub judice**, Latin for "under judgment", means that a particular case or matter is under trial or being considered by a judge or court; in other words, "pending adjudication."

that point. When accused of bias by Melero, he responded:
Dr. Finlay can say what he pleases concerning his work, while
Mr. Melero cannot comment about said work due to its private
nature. (López-Sánchez, 1987, p.105)

These are all the particulars of this incident recorded in Dr. López Sanchez's biography of Finlay. The matter, however, had not yet concluded. The minutes of the meeting found in the Annals indicate that, at the moment when passions were at their peak, Viñes spoke in defense of his friend, echoing the words of the President to the effect that Melero was out of order and adding an epigram: *"Facts can only be refuted by other facts"* (Annals of the Royal Academy, 1873, p.90). Viñes went on to state he had witnessed Finlay conducting the experiments, and the presence of alkali[61] in the rainwater tested was evident to him by observing how, as his cassock got wet in the falling rain, its acid stains had disappeared.

The President called for a vote, which went unanimously against Melero's accusations; Finlay had been vindicated. In a later session, on 28 September 1873 and without controversy, Finlay presented to the Academy a work entitled *"Transmission of cholera by means of water laden with specific principles (sic)"*, which appears in Volume 10 of the Academy Annals. In this paper, Finlay links water from the *"Royal Ditch"*, which supplied Havana with drinking water, to the transmission of cholera. In the ensuing discussion, and in support of Finlay's viewpoint, Viñes commented that the iron tubes carrying the Royal Ditch water would accumulate impurities over time.

These are just a few examples of the deep and abiding friendship between the two men, the meteorologist-priest and the physician, working in different fields, but joining forces in the service of society.

---

[61] Some authors define an alkali as a base that dissolves in water.

One can say that Viñes was not just a scientist, but also more specifically a scientist of Cuba. All his discoveries and his writing were rooted in Cuba's natural world. His awe at the beauty of the island's landscape and its flora is often expressed poetically in his writings. Witness what he has to say in this quote about the ubiquitous Royal Palm, *Roystonia Regia*, the national tree of Cuba:

> *How many palm groves I've visited, crisscrossing them in all directions, reading in them, as though it were written in indelible letters, the course of the hurricane… their graceful trunks laid on the ground and occasionally transported three, four and up to ten yards, that in their mute language appear to tell the visitor who watches them closely that, yes, there, from that precise point in the horizon came the fatal blow to which our stately and robust pillars succumbed.* (Viñes, 1877, p. 26)

Or in this other quote:

> *…the graceful and robust palm grows straight and upright amidst the vast surrounding plains, and reaches majestically toward the heavens with its sublime crown of fronds. The palm, projecting above all its surroundings, is always at the mercy of the wind, which one day gently sways its fronds to and fro and then, later, angrily attacks it and knocks it down, turning the palm into one of his first victims….* (Viñes, 1877, p.7)

In other parts of his work there are signs of his attunement to the traditional culture of the rural population of the island, and the respect he held for the close observations of weather phenomena that folk wisdom contained. This is a facet of Viñes' understanding of the relationship between science and society that had not appeared in his writings before. When discussing the electrical activity associated with hurricanes, Viñes writes these lines about a commonly held belief among rural folk in Cuba, stemming from his extensive contact with them:

This is a phenomenon often observed, that if at some point during the storm, one hears the rumble of thunder, or sees the flash of lightning, the end of the storm is close at hand. Thunderclaps and crowing roosters are the folk wisdom equivalent of a rising barometer that never fails them. (Viñes, 1877, p.190)

Viñes was known and esteemed widely among the common people, and for many years remembered as "the Father that taught about forecasting hurricanes." An elderly man, well into his 90s, who had grown up in the old Belén College neighborhood still vividly remembered him that way, when the author met him many years later. To this day, though all that knew him have passed on, Viñes' memory is still revered by many.

As happens with al celebrities, Viñes was occasionally the target of unfounded and even absurd rumors. An extreme example, since we know of no meteorologist that has ever been jailed as a result of a wrong forecast, was this note from the *La Lucha* newspaper from June 2nd 1890:

FATHER VIÑES JAILED

Last night the shocking news being discussed in cafés and theaters all over town was that Rev. Fr. Viñes, the distinguished Jesuit meteorologist, had been remanded to prison because he had not forecast, as was his duty, the recent heavy downpour that has caused so much damage. Some fools fell for this canard, which needless to say was entirely fictitious....

We can be sure that this item was received with much hilarity at the College, and must have seemed particularly curious to the person to whom, sadly, the expression "he was in prison and didn't even know it" did properly apply.

Another misconception about Viñes was a relatively common occurrence, even appearing in newspapers occasionally: referring to Viñes as an astronomer. The September 14, 1875 issue of *La Voz de Cuba*, for instance, twice refers to him that way, in addition to adding a curiously "crushing" way of referring to the recent hurricane Havana had experienced.

The illustrious astronomer of the Royal College of Belén, Fr. Benito Viñas (sic), whose observations and predictions about the hurricane that has recently passed on top of this city...

Though it is true that Viñes did some work in the field of Astronomy, and as has been mentioned, the Observatory counted with a good telescope, he was always known to fellow scientists as a meteorologist and not as an astronomer, though the general public could be forgiven for not knowing the difference between the two fields, given the common association of the words "observatory" and "astronomer."

Viñes was a member of several scientific societies in addition to being a Fellow of Havana's Royal Academy. Among these was the Scientific Society of Brussels, Belgium. We don't know if his membership in this society dates to his trip to Belgium in the 1880s. We do know, however, that he was appointed Corresponding Member of the German Meteorological Society on September 19th 1884. He also became a member of the Landowner Association, although it appears this was for no reason other than the contributions he made to agriculture with his meteorological research.

More than the awards and honors he received, surely the most valuable reward for Viñes was knowing that his forecasts had helped save the lives of countless men, women and children. One of those, the captain of an American ship wrote in the *Times Democrat* newspaper that the American government should supply Viñes with the latest scientific instruments, or better yet, a full observatory, if only as a token of recognition of the contributions he had made to American mariners plying the waters of the Caribbean and Atlantic. In Havana, the *Avisador Comercial* reprinted this article and added the following acerbic paragraph:

> *With only a tenth of the expenditures that are usually made in mounting an election of deputies that then go on to ignore the needs of the country, we would then be able to provide the best equipped observatory in the world to the eminent Jesuit and contribute to the honor of the country.* (El Avisador Comercial, 1890)

This was a sharp rebuke to the colonial government that allocated the resources to priorities that did not promote the advancement of the country and turned a deaf ear to repeated calls for of improvements in the country's meteorological service.

The Failed Forecast of 1888

It is also important to examine the biggest fiasco of Fr. Viñes' hurricane forecasting career: his failure to forecast the track of the strong hurricane that hit Cuba in September 1888. This hurricane followed a west-northwest track in the Atlantic after it formed, passing north of Puerto Rico and Hispaniola and reaching the Florida Straits along a track typical of other September storms he had studied. Once the hurricane reached waters north of Cuba, however, a strong high pressure system in the southern U.S. deflected it toward the west-southwest and brought it ashore along the north coast near the city of Caibarién.

Once ashore, the storm continued its trajectory along the former provinces of Las Villas, Matanzas, Havana and Pinar del Rio, reached the waters of the Yucatan Channel and went inland again over the Yucatan Peninsula. Current estimates put the strength of this hurricane as it reached Cuban coast in category 3 of the Saffir-Simpson scale.

The first advisory issued by Viñes was published on 4 September. It read:

Since this morning there have been subtle signs of a cyclone east of us, though we have as yet been unable to obtain a reliable estimate of the vortex location since it is far from us.

B. Viñes, S.J. (Diario de la Marina, 4 September 1888, morning edition)

Later that morning, Viñes received a telegram from Charles Ramsden in Santiago de Cuba, one of his more experienced collaborators. Ramsden alerted him to the presence of stormy weather in Saint Thomas and Puerto Rico. With this and other reports in hand, Viñes sent an updated press release to Havana newspapers, at noon on the same day:

The storm is located to the east of us, with a tendency to move slowly to our NE, with lowering barometer and a bad weather. This, along with the information we have received by telegram since this morning, appears to indicate that the storm track is coming much closer than previously thought... It is possible that we might experience gusty winds and showers this afternoon and tonight in the cities of Cárdenas, Matanzas and Havana. The vortex of the cyclone will likely pass through Key West and South Florida.

B. Viñes, S.J. (Diario de la Marina, 4 September 1888, afternoon edition)

It is clear that Viñes was expecting the hurricane to track significantly north of Cuba, with no more serious consequences to the island than strong winds and rough surf. Instead, the island's infrastructure and population were severely impacted by damaging winds and heavy rains for which they were unprepared. The most significant damage occurred as a result of coastal and inland flooding.

The town of Sagua la Grande, with a population of 12,000 at the time, took the brunt of the storm. The parish church served as a shelter for five hundred residents, including the soldiers from the local army garrison, since their barracks had been demolished by the winds. While the men inside the church tried in vain to prop up the church's three large doors, the winds eventually brought them down, causing collective panic among the refugees as the wind and rain engulfed them.

There were an estimated 600 deaths caused by this storm, along with approximately 10,000 people rendered homeless. Hundreds of people were said to have drowned in the nearby coastal settlement of La Boca, now known as Isabela de Sagua[62].

In this situation Viñes did not have sufficient synoptic data for the region north of Cuba for him to make a good forecast and had to rely almost exclusively of the cloud observations, barometric readings he made at the Observatory, together with the scant data he gleaned from telegrams he had received. It is also possible that he, guided by previous experience, had applied in an inflexible manner the first of his "laws":

GENERAL LAW CONCERNING THE TRACK OF HURRICANES IN THE ANTILLES

From its point of origin, the cyclone moves toward the fourth quadrant, heading more or less toward the west; later it begins

---

[62] Ramos, L. (2009): Huracanes. Desastres naturales en Cuba, Editorial Academia, La Habana

*to gradually bend northward, gradually recurving toward the
first quadrant, so that, in the final segment of its path it nor-
mally moves toward the NE or ENE.* (Viñes, 1895, p.31)

By the early morning hours of 5 September, as Havana experienced strengthening winds from the NW, gradually shifting to
N and E—the opposite direction from what he had anticipated—
Viñes realized that the hurricane was travelling over Cuba. Initially puzzled, Viñes eventually concluded that the reason for the
"abnormal" track of the cyclone was its interaction with another
cyclone located NE of Cuba, near the Florida peninsula.

Viñes apparently thought that the air circulation around cyclones behaved in a manner similar to electrical circuits; he assumed there would be some sort of repulsion between the "upper
air currents" of two adjacent cyclones. It would, therefore make
sense to him that the hurricane that was moving westward through
Cuba had to somehow evade a "collision" with the other cyclone
by shifting south of its "normal" trajectory, as he explained in the
4 September 1888 advisory.

A colorful sideshow to Viñes' failed forecast was provided by
an "amateur meteorologist" named Mariano Faquineto, a popular
snack and candy vendor that plied his trade in the plaza of the
nearby town of Guanabacoa, who delivered daily weather forecasts
to all within earshot. His accurate forecast of the 1888 hurricane,
much to the dismay of Viñes, caused this event to become known
to the people of Havana as the "Faquineto hurricane".

Benito Viñes could not always count on the energy required
by the demands of his rigorous work schedule. Coupled to his
naturally frail constitution, this contributed to a steady decline of
his health. Already by 1873, three years after his arrival in Cuba, we
see the first evidence of his physical deterioration. On October 12
of that year, a date when he was supposed to give an oral presentation to the Academy of Sciences, he instead sent his manuscript,
along with a note explaining his absence:

12 October 1873

Mr. President:

*By reason of my indisposition, and in view of the weather,
the Rev. Fr. Rector has deemed it prudent that I not go to the
Academy. By means of Fr. Ipiña, I am sending you a copy of
the little I have been able to accomplish so far, and that I would
have preferred to read in person.*

*Your humble servant
Benito Viñes S.J.*
(Archives of the Cuban Academy of Sciences, Benito Viñes
academic file)

The "indisposition" mentioned in this brief note appears fairly
serious, since it was reason enough for the Rector of the College
to forbid him from walking the few blocks to the Academy that
day. We have no sense of what kind of illness Viñes was dealing
with during that occasion. He was 36 years old at that time. Viñes
makes another reference to his health in the *Apuntes Relativos a
los Huracanes de las Antillas*…when he tells us how he was forced
to cut short a tour of the south part of Havana province in January
or February of 1877.

The way he expresses his fear of cold temperatures, we assume
he was concerned about catching pneumonia:

*I wanted to continue my investigations…but I must confess
that the persistent and unusually strong nortes made it almost
impossible for me to continue the trip without endangering my
health.* (Viñes, 1877, p.19)

Fr. Gutiérrez Lanza, in his *Apuntes Históricos…* affirms that
during the decade of the 80s, Viñes' health suffered several set-

backs, some very significant. It is possible that by then he might have been diagnosed with a pulmonary or cardiac affliction that would eventually lead to his death. Gutiérrez states that:

On many occasions, after a long and arduous day, he would still work all night until the illness, always undermining his strength, forced him to rest from his duties, if only for brief periods.... (Gutiérrez-Lanza, 1904, p. 19)

From the time his illness was diagnosed, Viñes began spending time at the Quinta La Asunción, a Jesuit-owned rest home located in the town of Luyanó, in the outskirts of Havana. Viñes would find sustenance going to that rural setting, perhaps recalling his childhood and adolescence in the Catalonian countryside, something that gave him an opportunity for quiet reflection. As time went on, he was forced to spend longer and longer periods there.

In 1892, the year before his death, Viñes experienced a significant setback that forced him to spend three months convalescing at La Asunción. In May of that year, the *Diario de La Marina* received the following letter that betrayed his melancholic state of mind:

Quinta La Asunción
Luyanó, 5th of May 1892

Finding myself in the country, convalescing from a long and dangerous illness, I find it impossible for now to continue my observations with the same zeal and favorable conditions of previous years.

I must, however, inform you that in my opinion there have been signs of the development of a cyclonic perturbation toward the WSW...

Benito Viñes, S.J.

This is evidence of the work ethic of a man unable to set aside his obligations even while dealing with a life-threatening illness. His sojourn in the countryside estate, probably prescribed—and enforced—by Finlay, was part of a treatment that included rest as its primary therapy. Viñes had probably ignored that therapeutic regimen whenever he was at home in the Observatory.

A little more than a month later, while still at *La Asunción*, Viñes sent three notes to the press in reference to another weak cyclonic event that crossed Cuba west of Havana between the 9th and 11th of June and produced widespread heavy rainfall and flooding in western and central Cuba, in the area between Las Villas and Havana provinces. The next report from Viñes is dated June 20th, sent from the Observatory after Viñes' return where he had returned after presumably having recovered sufficiently. He only had a year left to live. The last press release composed by Viñes was dated 29 October 1892, at 10:00 AM. In December of that year he spent his last Christmas among his brothers of the Jesuit Order.

Early in 1893, Viñes received an invitation from the organizing committee of the International Meteorological Congress, which was to be held in August 1893 in Chicago. This event was part of the World Columbian Exposition, more popularly remembered as the Chicago World's Fair, held to commemorate the 400th anniversary of Columbus' arrival in the New World, and to showcase the technical and scientific achievements of the United States as an emerging industrial power of the late nineteenth century, as well as Chicago's recovery from its devastating fire of 1871.

The exposition opened its doors to the public on May 1st 1893 and closed on the 30th of October of the same year. The International Meteorological Congress was held August 21-24 at the Exposition's monumental Palace of Fine Arts Building, which now houses the Chicago Museum of Science and Industry.

The International Meteorological Congress was meant to bring together the best and the brightest figures in the atmospheric sciences of the time. Presiding over the organizing committee for this event was Mark Harrington, who served as director of the newly renamed Weather Bureau from 1891 to 1895. Meteorologist H.C. Frankenfield[63] was vice chairman, and Professor Oliver Fassig[64] secretary of the committee. Other prominent member of the organizing committee included Cleveland Abbe and F.H. Bigelow.

The organizing committee sent out invitations to participate in the Congress to prominent meteorologists around the world in December 1892. The invitation received by Viñes, signed by Mark Harrington and Cleveland Abbe, asked Viñes to present the results of his work on the hurricanes of the Antilles in the session of the Congress devoted to theoretical meteorology.

Keeping in mind that Viñes had only until July 1893—a scant seven months—to prepare his presentation after receiving the invitation, all the while battling his final illness and suffering from failing vision, as some reports suggest, his stoic efforts in preparing his manuscript for the International Meteorological Congress are all the more remarkable.

This invitation prompted Viñes to use his last reservoir of strength to write the detailed treatise on hurricane circulation that became his scientific testament: *"Investigations Relating to the Circulation and Cyclonic Translation of Hurricanes of the Antilles,"* which summarizes all the work he had accomplished during his years in Cuba. By the time he was writing this work, his body was

---

[63] Harry Crawford Frankenfield (1862–1929) worked for the U.S. Weather Bureau and published several monographs, including reports on Mississippi floods and kite observations of vertical gradients of temperature, humidity and wind directions.

[63] Oliver Lanard Fassig (1860–1936) was an American meteorologist and climatologist who worked for the U.S. Weather Bureau initially as part of the Signal Corps of the United States War Department and later affiliated with the United States Department of Agriculture.

totally emaciated, "a mere skeleton covered by nerves and skin."

It is likely that the Government of Spain would have named Viñes as its official representative to the International Meteorological Congress, but knowing of his precarious health, named instead Fathers Federico Faura, S.J., and José Algué, both prominent meteorologists working at the Manila Observatory in the Philippines, to represent Spain at the Congress, since the Philippines were then under Spanish colonial rule.

Because of these circumstances, Benito Viñes' name at that conference became linked to Cuba and not to Spain, its colonial ruler, the sole affiliation listed under Viñes' name being the College of Belén in Havana, Cuba. Viñes' presentation is listed as contribution number 11 in the third session of "Section B—Theoretical Meteorology" of the Congress. In addition to his monograph, Viñes' *Cyclonoscope of the Antilles*[65] was shown in Gallery A at the Palace of Expositions as part of the collection of meteorological instruments shown in an exhibition.

In a memoir by the Spanish-Philippine meteorologists Faura and Algué, they write:

> *We take comfort in honoring his memory and treasuring this work that summarized the exhausting effort in meteorological research which he conducted over many years, a task that no doubt undermined his health. But it is with great satisfaction that we note that his work has been recognized as a landmark in the field by one of the most prominent North American meteorologists, Mr. Ferrell, who cites Fr. Viñes several times in his book "A Popular Treatise on the Winds" and by Everett Hayden, who was sent by the U.S. Government to Havana in 1888 to meet with Viñes to discuss topics related to the Hurricanes of the West Indies.*

---

[65] Faura, F. and J. Algué (1894): *La Meteorología en la Exposición Colombina de Chicago*, p. 61, Imprenta de Heinrich y Cia., Barcelona.

Viñes worked until the very end. He finished his monograph and mailed it to Chicago on the 21st of July. He died two days later, on Sunday the 23rd. His life ended at 10:45 AM from a "cerebral hemorrhage" according to the coroner's report.

Immediately after his death, his body was taken to the San Plácido chapel inside the chapel, where an honor guard was set up for the wake. By his casket that night stood in turn all his close collaborators: The rector of Belén, Father Benigno Iriarte, brother Alberdi, Father Gutiérrez-Lanza, and dozens of teachers and students of the College.

Early the following morning, filing by his casket went representatives of the major institutions of Havana, the press, and members of the Royal Academy of Sciences. The funeral mass began at 3:00 PM, led by the rector. At 4:30, the funeral cortege left the College. The route took it down Luz Street, then it turned on Monserrate, then onto Reina Street, Carlos III Boulevard, Zapata Street to finally arrive at the Colón necropolis, then located on the outskirts of the city. The people of Havana congregated in balconies and windows to bid farewell, in a spontaneous outpouring of grief.

Waiting at the cemetery were the principal aide to the governor, Col. Manuel Agudín, and the chief judge of the island. The Catholic Church was represented by the governor of the Diocese, Presbyter Casas; the Rev. Fr. Picabea, Dean of the Cathedral: and Rev. Fr. Iriarte, Rector of Belén College.

Among his friends were Carlos J. Finlay, brother Alberdi, the Gelats family, and many more. After the final words of farewell, an honor guard of the Anunciata Congregation carried the casket on their shoulders. Viñes was entombed at approximately 6:00 PM.

To avoid redundancy, suffice it to say that that the entire press corps of Cuba wrote extensive obituaries praising Viñes and all his achievements. At the Academy of Medical, Natural and Physical Sciences, Dr. Arístides Mestre read his *"Biographic Note about Fr. Benito Viñes,"* which incorporated parts of his last work, which Finlay had translated to English.

Viñes' tombstone is located near the cemetery chapel, some 100 meters from the main entrance. It is inscribed in Latin with the words *Sepulcrum Sodalius* and *Societatis Iesus*, the latter words indicating his membership in the Jesuit Order.

FIGURE 12. Viñes' tomb located in Havana's Colón Necropolis. (Photo by the author.)

Epilogue

A little while back I visited the hulking building that once housed the College of Belén of the Company of Jesus in Old Havana and that is now used for other purposes. Some of it has been restored, but the part once occupied by the Observatory is in ruins. I reached that part of the building by climbing stairs on the verge of collapse, crossing rooms with sagging floors and crumbling walls, finally arriving at the old Observatory. I was alone in that deserted corner of the building, feeling the afternoon breeze blowing through the openings that used to frame windows that looked out on the bay and Old Havana. There, blurring my vision, I could almost make out a small man, in black cassock wearing small round spectacles, that without looking at me, said: Can you see those clouds? They are cirrus. I've been waiting for them since yesterday.

Works by Fr. Benito Viñes

Viñes, B., 1870: Los Huracanes del 7 y 19 de octubre de 1870 (The Hurricanes of 7 and 19 October 1870). Manuscript available at the Library of the Cuban Academy of Sciences.

Viñes, B., 1872: Perturbaciones magnéticas y aurora boreal del 4 de febrero de 1872 (Magnetic perturbations and the aurora borealis of 4 February 1872). *Annals of the Royal Academy*, v. 9. [The full Spanish title of the *Annals of the Royal Academy* is *Anales de la Real Academia de Ciencias Médicas, Físicas y Naturales de La Habana*. The short English title is used hereafter in the bibliography.]

Viñes, B., 1873: Perturbaciones magnéticas en relación con los demás elementos meteorológicos durante los meses de junio a octubre (Magnetic perturbations in relation to other meteorological elements during the months of June through October). *Annals of the Royal Academy*, v. 10.

Viñes, B., 1873: Temporal del 6 de octubre de 1873 (The storm of 6 October 1873). *Annals of the Royal Academy*, v. 10.

Viñes, B., 1873: Marcha regular o periódica e irregular del barómetro en La Habana desde 1858–1871 inclusive (Regular and periodic and irregular march of the barometer in Havana from 1858 through 1871). *Annals of the Royal Academy*, v. 11.

Viñes, B., 1877: *Apuntes Relativos a los Huracanes de las Antillas en Septiembre y Octubre de 1875 y 76 (Notes concerning the hurricanes of the Antilles during September and October 1875 and 76)*. "El Iris" Typography and Stationery, Havana.

Viñes, B., 1870: Huracanes del 7 y 19 de Octubre de 1870 (Hurricanes of 7 and 19 October 1870). Magnetic and Meteorological Observatory of the College of Belén (no other publication information available).

Viñes, B., 1880: "Excursión a Vueltabajo. Realizada por el R.P. Benito Viñes, S.J. y el Sr. D. Pedro Salterain" ("Field Trip to Vueltabajo by Rev. Fr. Benito Viñes, S.J., and Don Pedro Salterain"). Special supplement to *El Triunfo* newspaper.

Viñes, B., 1881: Las perturbaciones magnéticas en relación con los Nortes y principales cambios atmosféricos (Magnetic perturbations in relation with "nortes" and principal atmospheric fluctuations). *Magnetic and Meteorological Observations made at the Royal College of Belén, Havana.*

Viñes, B., 1882: "Las Tempestades de Cuba" (The Storms of Cuba). *Annals of the Royal Academy,* v. 24.

Viñes, B., 1883: Determinaciones absolutas de la declinación, inclinación y fuerza horizontal magnética terrestre hechas en en Observatorio del Colegio de Belén en La Habana (Measurements of the absolute value of the declination, inclination and strength of the terrestrial magnetic force made at the Observatory of the College of Belén). *Annals of the Royal Academy,* v. 24.

Viñes, B., 1883: Observación del Paso de Venus, hecha en el Real Colegio de Belén de La Habana, el 6 de diciembre de 1882 (Observations of the Transit of Venus, made at the Royal College of Belén in Havana, on 6 December 1882). *Annals of the Royal Academy,* v. 24 pp. 310-321.

Viñes, B., 1888: *Ciclonoscopio de las Antillas (Cyclonoscope of the Antilles).* Ricardo B. Caballero Lithographers, Havana.

Viñes, B., 1888: Trayectoria anormal del desastroso ciclón del 4 al 5 de septiembre de 1888 (Anomalous trajectory of the disastrous hurricane of 4 and 5 September 1888). *Magnetic and Meteorological Observations made at the Royal College of Belén, Havana, 1888 (Appendix).*

Viñes, B., 1895: *Investigaciones Relativas a la Circulación y Traslación Ciclónica en Los Huracanes de Las Antillas (Investigations concerning the cyclonic circulation and trajectory of hurricanes in the Antilles),* 1st ed. The Pulido y Díaz Press of the Commercial Advertiser, Havana.

References

Academia de Ciencias de Cuba y Academia de Ciencias de la URSS *(Academies of Science of Cuba and the USSR)*, 1970: *Atlas Nacional de Cuba. (National Atlas of Cuba)*. National Directorate of Geodesy and Cartography, Council of Ministers of the USSR, Moscow, 132 pp.

Archivo Histórico de la Academia de Ciencias de Cuba (Historical Archive of the Cuban Academy of Sciences), 1873–1893: Expediente de Benito Viñes en la Real Academic de Ciencias Médicas, Físicas y Naturales de La Habana (Benito Viñes' personnel file at the Royal Academy of Medical, Physical and Natural Sciences of Havana).

Archivo Histórico del Inisituto de Meteorología del Ministerio de Ciencia, Tecnología y Medio Ambiente, 1881–1888: Colección de notas de prensa del Observatorio del Colegio de Belén (Collection of press clippings from the Observatory of the College of Belén kept at the Historical Archives of the Institute of Meteorology).

Archivo Nacional de Cuba (National Archives of Cuba), 1861: Royal Decrees and Regulations, Vol. 187, No. 10496.

Atlas Ilustrado Labor, Editorial Labor, Barcelona (Labor Illustrated Atlas, Labor Press, Barcelona), 1971.

Aurora del Yumurí, 1870: Revista de los desastres ocurridos en los días 7 y 8 de octubre de 1870 (Highlight of the disasters that took place on October 7 and 8, 1870). Matanzas, 14 October.

Committee on Clouds and Cloud Formations of the Weather Bureau. United States Department of Commerce (1945). Cloud Code. U.S. Government Printing Press.

Constantino, C. E., 1943: *Meteorología Descriptiva (Descriptive Meteorology)*. 1st ed. Sopena Press, 487 pp.

de la Torre, J. M., 1857: *Lo Que Fuimos y Lo Que Somos ó La Haban Antigua y Moderna (What We Were and What We Are, or Old and New Havana)*. Spencer & Co. Press, Havana, 187 pp.

de Arrate, J. M. Felix, 1964: *Llave del Nuevo Mundo, Antemural de las Indias Occidentales (Key to the New World, Threshold of the West Indies)*. Cuban Commission of UNESCO, Havana, 75 pp.

de la Torriente-Brau, Z., 1975: Indice Analítico de los *Anales de la Real Academia de Ciencias Médicas, Físicas y Naturales de la Habana* (Analytical Index of the *Annals of the Royal Academy of Medical, Physical and Natural Sciences of Havana)*. Cuban Academy of Sciences, 175 pp.

Dunn, G. E., et al., 1969: *Atlantic Hurricanes*. (Cuban Institute of Meteorology, Academy of Sciences, 377 pp.

Eguren, G., 1989: La fidelísima Habana (The most faithful Havana). Cuban Letters Press, 436 pp.

Finlay de Barrés, C. J., 1873: Alcalinidad atmosférica (Atmospheric alkalinity). *Annals of the Royal Academy,* La Antilla Press, v. 10.

———, 1873: Transmisión del cólera mediante el agua de lluvia cargada con pricipios específicos (Transmission of cholera by means of contaminated rainwater). *Annals of the Royal Academy,* La Antilla Press, v. 10.

Flammarion, C., 1873: *L'Atmosphere (The Atmosphere)*. 2nd ed. Hachette & Co. Press, Paris, 120 pp.

Gangoiti, S. J., 1895: Prólogo (Prologue). *Investigaciones Relativas a la Circulación y Translación Ciclónica en los Huracanes de las Antillas* (Investigations concerning the cyclonic circulation and tracks of the hurricanes of the Antilles), B. Viñes, Avisador Comercial Press, Havana, 5–13.

Gutiérrez-Lanza, M., 1904: *Album Conmemorativo del Cincuenta Aniversario de La Fundación en La Habana del Colegio Belén de la Compañía de Jesús* (Commemorative Album of the Fiftieth Anniversary of the Founding in Havana of the College of Belén of the Company of Jesus). The Commercial Advertiser Press, Havana, 410 pp.

————, 1904: Apuntes Históricos acerca del Colegio de Belén. La Habana. (Historical Notes about Belén College, Havana).

————, 1927: *Génesis y Evolución del Huracán del 20 de Octubre de 1926 y Catálogo de Ciclones de la Isla de Cuba, 1865–1926. (Genesis and Evolution of the Hurricane of 20 October 1926 and Catalogue of Cyclones of the Island of Cuba, 1865–1926).* Dorrbecker Press, Havana, 51 pp.

————, 1934: Ciclones que han pasado por la Isla de Cuba (Cyclones that have traversed the Island of Cuba). Observatorio del Colegio de Belén, Cultural, S. A. 23 pp.

————, 1936: El Padre Benito Viñes S. J. y su Obra Científico-Humanitaria al frente del Observatorio del Colegio de Belén. Conferencia leída en el acto publico de exposición de libros sobre el padre Benito Viñes, S. J. celebrado en la bibiloteca municipal de La Habana el día 26 de Julio de 1936. Publicaciones de la Biblioteca Municipal de La Habana (Father Benito Viñes, S. J., and his scientific-humanitarian works as head of the College of Belén's Observatory. Lecture presented in the public presentation of books about Father Benito Viñes held at the Havana Municipal Library on July 26, 1936). Publications of the Municipal Library of Havana, Series "B" Popular Culture, No. 3, 30 pp.

————, 1942: El Padre Benito Viñes, S.J. *Figuras Cubanas de la Investigación Científica* (Prominent Cuban Scientific Researchers). Publications of the Havana Ateneum, Vol. II, 112-140.

Herrick, F., 1927: *The Man Who Named The Clouds.* Tycos-Rochester, v. 17.

Leal, E., 1988: *La Habana Antigua y Moderna (Old and Modern Havana).* Cuban Letters Press, 123 pp.

López, A., 1957: Contribución a una biografía completa del padre Benito Viñes (Contribution to a complete biography of Fr. Benito Viñes). Taller de Artes Gráficas de los hermanos Bedia, Santander, 96 pp.

López Sánchez, J., 1987: *Finlay: El hombre y la verdad científica (Finlay: The man and scientific truth)*. Scientific-Technical Publishers, Havana, 76 pp.

Millás, J. C., 1938: Notas históricas sobre el huracán del 4 y 5 de octubre de 1844 (Historical notes on the 4 and 5 October 1844 hurricane). *Boletín del Observatorio Nacional (Bulletin of the National Observatory)*, epoch III, Vol. III, no. 1, Havana.

Observatorio del Colegio de Belén, 1870: Observaciones magnéticas y meteorologicas hechas por alumnos del Colegio de Belén de la Compañía de Jesús en La Habana. Año meteorológico del 30 de noviembre de 1868 al 30 de noviembre de 1869 (Magnetic and meteorological observations made by students of the College of Belén of the Society of Jesus in Havana. Meteorological year from 30 November 1868 to 30 Novermber 1869). Religious Publishers and Bookstore, Havana.

————, 1904: *Observaciones Magnéticas y Meteorológicas del Colegio de Belén de la Compañía de Jesús en La Habana (Magnetic and Meteorological Observations of the College of Belén of the Company of Jesus in Havana)*. Avisador Comercial Press, xxx pp.

Ortiz, R., 1979: Andrés Poey: Precursor de la meteorología científica en Cuba (Andrés Poey: Precursor of scientific meteorology in Cuba). *Conferencias y Estudios de Historia y Organización de la Ciencia (Presentations and Studies of the History and Structure of Science)*, No. 13, 15–16.

Piddington, H., 1859: *Guide Du Marin Sur La Loi Des Tempetes (Mariner's Guide to the Law of Storms)*. 2nd ed. Mallet-Bachelier Printers and Bookstore, 332 pp.

Rodrígues Ramírez, M., 1956: Cronología clasificada de los ciclones que han azotado a Cuba desde 1800 hasta 1956 (157 años) (Chronology of cyclones that hit Cuba for the period 1800 to 1956 (157 years). *Revista de la A.C.A.M. (Reviews of A.C.A.M.)*, Vol. II, No. 4 (December), 1–11.

Roman, C., et al., 1990: *Almanaque Mundial 1991 (World Almanac 1991)*. University of Miami Press, 640 pp.

Sarasola, S., 1926: *Los Huracanes de las Antillas (The Hurricanes of the Antilles)*. Bruno del Amo, Ed., La Moderna Poesía, 171 pp.

Secchi, A., 1875: *Le Soleil (The Sun)*. Gauthier-Villars Publishing and Bookstore, Paris, 428 pp.

Shepherd, T., 1956: *Historical Atlas*. 8th ed. Barnes and Noble, 115 pp.

Sociedades Bíblicas Unidas, 1989: Dios habla hoy: La Biblia (God Speaks Today: The Bible). Sociedades Bíblicas Unidas, 305 pp.

Tannehil, I. R., 1938: *Hurricanes: Their Nature and History*. Princeton University Press, 266 pp.

Udías, A., 1993: Los jesuítas y la meteorología (The Jesuits and meteorology). Paper presented in June 1993 during the Symposium on Tropical Cyclones "Fr. Benito Viñes In Memoriam", Havana.

U.S. Hydrographic Office (various years): Pilot charts of the Central American Waters. (various years). Available from the U.S. Hydrographic Office.

Valero, M., 1989: El Observatorio del Colegio de Belén en el siglo XIX (The College of Belén Observatory in the nineteenth century) *CEHOC Annuary*, no. 1, Academia Press, Havana.

Newspapers Consulted

Diario de la Marina, Havana. *(Translator's note:* Diario de la Marina, founded in 1832 and closed after the 1959 Revolution, was a conservative-leaning newspaper with the largest circulation in Cuba.)

El Avisador Comercial (The Commercial Adviser), Havana.

El Bien Público (The Public Good), Santiago de Cuba.

El Boletín Mercantil (The Merchants' Bulletin), San Juan, Puerto Rico.

El Triunfo (The Triumph), La Habana.

La Iberia (Iberia), Havana.

La Lucha (The Struggle), Havana.

La Voz de Cuba (The Voice of Cuba), Havana.

Index